Lumpy Skin Disease

Eeva S. M. Tuppurainen • Shawn Babiuk •
Eyal Klement

Lumpy Skin Disease

 Springer

Eeva S. M. Tuppurainen
Aldershot, Hampshire, UK

Shawn Babiuk
Canadian Food Inspection Agency
National Centre for Foreign Animal Disease
Winnipeg, Manitoba, Canada

University of Manitoba
Winnipeg, Manitoba, Canada

Eyal Klement
Koret School of Veterinary Medicine
The Hebrew University
Rehovot, Israel

ISBN 978-3-030-06427-3   ISBN 978-3-319-92411-3   (eBook)
https://doi.org/10.1007/978-3-319-92411-3

This Springer imprint is published by the registered company Springer International Publishing AG part of Springer Nature.
The registered company address is: Gewerbestrasse 11, 6330 Cham, Switzerland

Contents

1   **Introduction to Lumpy Skin Disease**   1

2   **General Description of Lumpy Skin Disease**   3
     References   5

3   **Economic Impact of Lumpy Skin Disease**   7
     3.1   Direct Impact of Clinical Disease   7
     3.2   Indirect Economic Impact of LSD   8
     References   9

4   **Geographic Distribution of Lumpy Skin Disease**   11
     References   13

5   **Current Legislation and Trade Recommendations**   15
     References   17

Part I   Lumpy Skin Disease Virus

6   **Taxonomy**   21
     References   23

7   **Morphology**   25
     References   27

8   **Genome**   29
     References   35

9   **Replication in a Host**   37
     References   39

10   **Propagation of the Virus In Vitro**   41
     References   43

11   **Persistence and Stability of the Virus**   45
     References   45

12   **Immunity**   47
     References   50

13 Epidemiology and Transmission   53
    13.1   Transmission Modes of LSDV   53
    13.2   Potential Vectors of Lumpy Skin Disease Virus   54
    13.3   Direct Transmission of Lumpy Skin Disease   55
    13.4   Transmission Via Subclinically Infected Cattle   56
    13.5   Spread of Lumpy Skin Disease   56
    13.6   Seasonality   57
    13.7   Geographical Risk Factors   57
    13.8   Risk Factors in the Herd Level   58
    13.9   Risk Factors in the Animal Level   58
    13.10   Further Suggested Studies   59
    References   60

Part II   Early Detection of Lumpy Skin Disease, Diagnostic Tools and Treatment

14 Clinical Signs   65
    References   69

15 Sample Collection and Transport   71
    References   72

16 Diagnostic Tools   73
    References   77

17 Treatment of Lumpy Skin Disease   81

Part III   Control and Eradication

18 Vaccines Against LSD and Vaccination Strategies   85
    18.1   Vaccination Strategies   90
    References   91

19 Slaughter of Infected and In-Contact Animals   95
    References   96

20 Animal Movement Control and Quarantine   97
    References   98

21 Vector Surveillance and Control   99
    21.1   Vector Surveillance   99
    21.2   Vector Control   100
    References   102

22 Decontamination and Disinfection   103
    References   105

23 Active and Passive Surveillance   107
    References   109

Introduction to Lumpy Skin Disease

Eeva S. M. Tuppurainen

Lumpy skin disease (LSD) is a high-impact transboundary pox disease of cattle and Asian water buffalo caused by a lumpy skin disease virus (LSDV). The disease is categorized on the list of notifiable diseases by the World Organization for Animal Health (OIE) due to its capability of rapid transboundary spread and to cause substantial cattle production losses. The OIE sets up recommendations for the export and trade of live cattle and their products from affected countries. Within the European Union (EU), notification, disease control, vaccination, cattle movements and trade within and between the EU member states are strictly regulated by different directives and implementing decisions.

Lumpy skin disease is widely spread in Africa and in the Middle East. In 2013 the first LSD outbreaks were reported in Turkey from where it swiftly spreads to the northern part of Cyprus, Southeast Europe and the northern Caucasus region. Currently, those countries neighbouring the affected areas in Europe and Asia are facing a risk of incursion of LSD into their territories.

An incursion of LSD into a previously disease-free country causes severe losses for all sections of the cattle farming industry. Decrease in milk and meat production, abortions, fertility problems, damaged skins and hides as well as death or culling of sick cattle comprise the direct losses. Furthermore, indirect losses follow the cattle movement and trade restrictions. Livelihoods of poor smallholders and backyard farmers are most severely affected, but similarly incursion of LSD into an intensive beef or dairy cattle production unit causes extensive economic losses to the producers as well as environmental issues in case a total or even modified stamping-out policy is in place.

Usually the very first suspicion of LSD is raised when several febrile animals with highly characteristic skin nodules, eye discharge and enlarged lymph nodes are detected by cattle owners. Dairy cattle are daily monitored, but the virus may sometimes circulate for weeks in free-ranging beef cattle herds before detected, allowing plenty of time for vectors to become infected and spread the virus to naïve susceptible herds in the region.

E. S. M. Tuppurainen et al., *Lumpy Skin Disease*,
https://doi.org/10.1007/978-3-319-92411-3_1

Within the currently affected regions, mass vaccination with sufficient coverage is fundamental for halting the spread of a vector-borne LSDV supported by the other control measures. To date, none of the affected countries has been able to permanently eradicate disease, once it has got a foothold in their territories. However, the effectiveness of the total stamping-out measure is likely to vary depending on the region and cattle farming practices. In case outbreaks are detected in a very early stage, epidemiological unit sizes are small and cattle movements can be properly controlled, a total stamping-out measure would probably stop the spread of LSD without vaccination.

Ideally, in a face of an outbreak, the vaccination of the whole cattle population in a country or zone should be carried out within the shortest possible time frame. Depending on the local cattle farming practices, terrain, roads and available means of transport, vaccination campaigns are usually very time-consuming and laborious, stretching the veterinary services up to their limits. Despite of the challenges, the efficacy of a well-organized vaccination campaign, using effective vaccines, has been recently demonstrated by the extremely successful control of LSD in Southeast Europe and the Western Balkan. However, the costs of mass vaccination are a substantial economic burden, considering the fact that in 2013–2017 the majority of the affected countries had to operate with limited financial resources.

Before 2012 LSD was a largely neglected cattle disease and raised only a limited research interest outside the endemic regions. It was believed that LSD was not going to spread out from Africa or in case it would happen, the disease could be easily contained. Thus, LSD was not considered as a disease which required investment in research. The lack of general interest leads to limited scientific research on immunity, epidemiology, transmission, vectors, viral characteristics, diagnostic tools, veterinary treatments and prophylactic agents. Triggered by the recent spread of LSD, the situation has fundamentally changed, and the availability of funding has initiated novel research and international collaborations, producing much needed and long-overdue scientific data for improved control of the disease.

The aim of this book is to provide an overview on the essential aspects of LSD based on the most recent research data.

General Description of Lumpy Skin Disease

2

Eeva S. M. Tuppurainen

Lumpy skin disease (LSD) is endemic across most of the African continent, and since 2012 it has spread widely within the Middle East, Southeast Europe and the northern Caucasus. Lumpy skin disease virus (LSDV) shares the genus *Capripoxvirus* (CaPV) within the family *Poxviridae* (Buller et al. 2005) with sheeppox virus (SPPV) and goatpox virus (GTPV).

The World Organization for Animal Health (OIE) provides recommendations for the international trade standards in the Terrestrial Animal Health Code, Infection with Lumpy Skin Disease Virus, Chap. 11.9, and for diagnostic assays and vaccines in Lumpy Skin Disease, Chap. 2.4.13 of the Manual of Diagnostic Tests and Vaccines for Terrestrial Animals. The most recent version of the LSD chapter was adopted in May 2017.

European Union (EU) regulates the disease notification (82/894/EEC of 21 December 1982), intra-community trade of live animals and their products (90/425/EEC of 26 June 1990), as well as control and eradication measures applied in case of an outbreak within the EU member states (92/119/EEC of 17 December 1992).

Infection with LSDV causes clinical signs in cattle and domestic (Asian) water buffalo. Although some wild ruminants are known to be susceptible, the role of wildlife in the transmission of the disease is not clear.

Sudden appearance of fever and skin lesions in several animals in the herd is a characteristic finding for LSD. Sometimes infected animals develop only a few lumps, but in severe cases skin nodules may cover the entire body. Affected cattle show ocular and nasal discharge, and ulcerative lesions may be detected in the oral, nasal and ocular mucous membranes. Typically, infected cattle show clearly enlarged subscapular and precrural lymph nodes. Swellings, caused by oedema, can be detected in the dewlap and legs, causing lameness (Haig 1957; Weiss 1968). Secondary bacterial skin infections, mastitis and especially pneumonia can be so severe that they may lead to death or euthanasia of affected animals.

An incursion of LSDV to a previously disease-free country is usually associated with either legal or illegal introduction of cattle, originating from affected region

© Springer International Publishing AG, part of Springer Nature 2018
E. S. M. Tuppurainen et al., *Lumpy Skin Disease*,
https://doi.org/10.1007/978-3-319-92411-3_2

(Jarullah 2015; Ince et al. 2016). After arrival of clinically or subclinically infected animal(s), blood-feeding vectors (Weiss 1968; Kitching and Mellor 1986; Chihota et al. 2001; Lubinga et al. 2013; Tuppurainen et al. 2013a, b) further disseminate the virus to naïve animals within their flying distance or when cattle comes to close contact with each other. Although there is no absolutely safe season for LSD, the outbreaks usually occur in the spring, summer and autumn when the temperature and humidity are ideal for biting and blood-feeding vectors.

So far, only mechanical transmission by vectors has been demonstrated. Further studies are, however, required to exclude the possibility of much more effective biological vector transmission modes. To a lesser extent, transmission may occur in the absence of vectors by direct or indirect contact or via contaminated feed or water (Haig 1957; Weiss 1968). In addition, seminal transmission has been experimentally demonstrated (Annandale et al. 2013).

Severity of the clinical disease depends on the virulence of the virus and the factors affecting the susceptibility of the host, such as immunity, age and breed of cattle. Natural resistance to infection, not associated with immunity, is believed to occur in cattle (Weiss 1968) although this old statement needs to be experimentally re-evaluated by using modern highly sensitive molecular tests. Only half of experimentally infected cattle are likely to develop generalized skin lesions. The remaining animals either show only a localized swelling at the inoculation site of the challenge virus or no other clinical signs apart from a mild fever reaction (Weiss 1968; Tuppurainen et al. 2005; Osuagwuh et al. 2007; Annandale et al. 2010). Infected cattle do not remain asymptomatic carriers. Immunity against LSDV is both humoral and cell mediated. The levels of LSD-specific antibodies are highest following infection and decrease over time.

LSDV is a very stable virus which survives well in the farm premises and surroundings. A total ban or efficient control of animal movements is often challenging in affected regions. To date it has not been possible to halt the spread of the vector-borne LSD by culling of all infected and in-contact animals without using mass vaccinations.

In recently affected regions in Southeast Europe, large-scale immunization of cattle population has provided to be the most effective tool to control the spread of the disease. Supporting measures such as movement controls, stamping-out, disinfection and farm biosecurity measures are of high importance in combatting the spread. However, selection of the most feasible control and eradication policy, with or without vaccination, varies between countries and regions, depending on several factors and circumstances, such as the climate, the local farming practices and the epidemiological unit size.

Vaccines containing live attenuated LSDV are superior compared to sheeppox virus containing vaccines. To some degree, a cross protection seems to be in place across all members of the genus *Capripoxvirus*. However, there are differences in the level of protection provided by the different CaPV vaccines in cattle, and therefore selection of the vaccine should always be based on demonstrated efficacy in controlled studies.

LSDV is very stable in clinical samples. For primary detection of the LSD virus, several validated and highly sensitive capripox group-specific PCR methods have been described. Molecular diagnostic tools are also available to differentiate an attenuated vaccine virus from the virulent field strain as well as to differentiate between different members of the *Capripoxvirus* genus. For long serological tools were limited to neutralisation assays. Currently also capripox-specific enzyme-linked immunosorbent assay (ELISA) has become commercially available.

References

Annandale CH, Irons PC, Bagla VP et al (2010) Sites of persistence of lumpy skin disease virus in the genital tract of experimentally infected bulls. Reprod Domest Anim 45:250–255. https://doi.org/10.1111/j.1439-0531.2008.01274.x

Annandale CH, Holm DE, Ebersohn K, Venter EH (2013) Seminal transmission of lumpy skin disease virus in heifers. Transbound Emerg Dis 61:443–448. https://doi.org/10.1111/tbed.12045

Buller RM, Arif BM, Black DN et al (2005) Poxviridae. In: Fauquet CM, Mayo MA, Maniloff J et al (eds) Virus taxonomy: eight report of the international committee on the taxonomy of viruses. Elsevier Academic Press, Oxford, pp 117–133

Chihota CM, Rennie LF, Kitching RP, Mellor PS (2001) Mechanical transmission of lumpy skin disease virus by *Aedes aegypti* (Diptera: Culicidae). Epidemiol Infect 126:317–321

Haig DA (1957) Lumpy skin disease. Bull Epizoot Dis Africa 5:421–430

Ince ÖB, Çakir S, Dereli MA (2016) Risk analysis of lumpy skin disease in Turkey. Indian J Anim Res 50:1013–1017. https://doi.org/10.18805/ijar.9370

Jarullah BA (2015) Incidence of lumpy skin disease among Iraqi cattle in Waset Governorate, Iraq republic. Mater Methods 3:936–939

Kitching RP, Mellor PS (1986) Insect transmission of capripoxvirus. Res Vet Sci 40:255–258

Lubinga JC, Tuppurainen ESM, Stoltsz WH et al (2013) Detection of lumpy skin disease virus in saliva of ticks fed on lumpy skin disease virus-infected cattle. Exp Appl Acarol 61:129–138. https://doi.org/10.1007/s10493-013-9679-5

Osuagwuh UI, Bagla V, Venter EH et al (2007) Absence of lumpy skin disease virus in semen of vaccinated bulls following vaccination and subsequent experimental infection. Vaccine 25:2238–2243. https://doi.org/10.1016/j.vaccine.2006.12.010

Tuppurainen ESM, Venter EH, Coetzer JAW (2005) The detection of lumpy skin disease virus in samples of experimentally infected cattle using different diagnostic techniques. Onderstepoort J Vet Res 72:153–164

Tuppurainen ESM, Lubinga JC, Stoltsz WH et al (2013a) Mechanical transmission of lumpy skin disease virus by *Rhipicephalus appendiculatus* male ticks. Epidemiol Infect 141:425–430. https://doi.org/10.1017/s0950268812000805

Tuppurainen ESM, Lubinga JC, Stoltsz WH et al (2013b) Evidence of vertical transmission of lumpy skin disease virus in *Rhipicephalus decoloratus* ticks. Ticks Tick Borne Dis 4:329–333. https://doi.org/10.1016/j.ttbdis.2013.01.006

Weiss KE (1968) Lumpy skin disease virus. Virol Monogr 3:111–131

Economic Impact of Lumpy Skin Disease

3

Eyal Klement

3.1 Direct Impact of Clinical Disease

The incidence of LSD is the first factor which determines its direct economic impact. This depends on the abundance of vectors, the susceptibility of the host and the use of preventive measures (Gari et al. 2011). It can reach even 85% in an affected herd if no preventive measures are applied (Tuppurainen and Oura 2011). Case fatality is also an important factor, influencing the economic impact of a disease. However, accurate estimation of case fatality is very difficult to provide as in most of the developed countries, sick animals are culled and in developing countries, the exact pathological reason for natural animal death is not always provided. In an interview-based study performed in Ethiopia, a case fatality of 9.3% and 21.9% was reported in zebu and crossbred/Holstein Friesian cattle, respectively (Gari et al. 2011). In Albania the case fatality reported was 5.8% (364/6235) (EFSA AHAW Panel 2015). Turkish researchers reported a much higher case fatality in cattle in Turkey, reaching 54.8% in Holstein cattle (Sevik and Dogan 2017). Mortality (which is the product of incidence and case fatality) usually does not exceed 1%–3% for LSD in most situations (Tuppurainen and Oura 2011).

Apart from mortality, economic impacts are also caused by production losses, mainly loss of milk production, reduction in the percentage of animals slaughtered for food (offtake rate) in beef cattle and fewer days of draft animal power. In Ethiopia, the percentage of milk production lost was estimated to be 1.5% (51 L per lactation of an affected cow) and 3% (312 L per lactation of an affected cow) for zebu and crossbred/Holstein Friesian cattle, respectively (Gari et al. 2011). In Turkey, average loss of 159 L per lactation was calculated for a surviving affected cow (Sevik and Dogan 2017). It should be taken into account, however, that in both studies it was not detailed how exactly milk loss was estimated.

Among beef cattle, morbidity-related losses from LSD interfere with normal herd dynamics. Examples include lower reproductive rates among breeding stock and reduced weight gain in finished stock for the market, because of long-term morbidity.

© Springer International Publishing AG, part of Springer Nature 2018
E. S. M. Tuppurainen et al., *Lumpy Skin Disease*,
https://doi.org/10.1007/978-3-319-92411-3_3

The annual reduction in offtake rates in beef cattle in Ethiopia was estimated at 1.2% and 6.2% for zebu and crossbred/Holstein Friesian cattle, respectively (Gari et al. 2011). Abortions and infertility were also mentioned in some studies (Sevik and Dogan 2017; Tuppurainen and Oura 2011). However, their direct association with LSD is still not clear and is a subject for further studies. Another important source of losses include damage to hides as a result of scars from LSD (Tuppurainen and Oura 2011).

Lastly, in the highlands of Ethiopia, livestock are also used as draft animals for ploughing during the cropping season. Gari et al. (2011) estimated that 16 days of draft power are lost for an LSD-affected ox that survives. The impact of this effect depends upon the timing in which the disease occurs, as the cropping season in Ethiopia lasts only 2 months (Gari et al. 2011).

3.2   Indirect Economic Impact of LSD

The economic impact of LSD may reach beyond the individual affected farms, for example, through broader impacts on consumers and other members of society (e.g. taxpayers) or employment and income in affected communities or even international trade. Such an analysis is beyond the scope of this chapter but would include not only the financial losses accrued to individual farms afflicted with LSD but also these downstream or in some cases external costs to other members of society (Thrusfield 2005).

For example, mortality of cattle due to the disease itself or due to preventive culling might result in a reduced availability of milk and meat, which in turn could increase the price that consumers pay for these products. This represents a transfer of wealth from consumers to the remaining producers, as opposed to a loss. Nonetheless, it represents a downstream effect that is felt beyond the farm gate.

Another cost of LSD arises from control efforts, such as quarantines of farms neighbouring an infected herd. This control effort reduces the ability of people to freely move their livestock, sometimes to critical grazing areas or to markets. Such quarantines may be economically efficient—generating more benefits than costs—but some of those costs may be incurred by livestock owners whose herds are not infected. These costs should be recognized and accounted for properly in disease control decisions (Peck and Bruce 2017).

Other expenditures on control efforts may include vaccination, medication, human labour and, above all, stamping out. Some of these control costs are incurred by the owners of infected herds, but in some countries the government (i.e. taxpayers) pays for them. It is difficult to provide a general cost estimate of these control measures because there is large variability in the set of measures applied in different countries and in the value of cattle that may be culled for control purposes. Such variability was observed during the large LSD epidemic that occurred in Europe during 2015–2016. For example, Albania reported a total of 6235 affected animals through 2016, but only a few hundred died or were culled as a result of the disease. In comparison, Greece reported only 1000 affected animals. However, they enacted a policy of total stamping out of all animals on affected farms, without differentiating between vaccinated and non-vaccinated herds. Therefore, about 12,000 cattle were culled in Greece as a result of the LSD epidemic (EFSA AHAW Panel 2015; Agianniotaki et al. 2017). These differences in control policies arise from differences among individual countries' legislation (Peck and Bruce 2017).

International trade limitations can also be a major cause of loss during LSD outbreaks. These also differ significantly among countries. During an Ethiopian outbreak, rejection of affected bulls accounted for more than twice the losses associated with mortality (Alemayehu et al. 2013). In EU countries, where trade restrictions are stricter, the potential losses from trade suspension due to an LSD outbreak are expected to be even higher, relative to mortality losses. Trade suspension may apply not only to live animals, or meat and dairy, but also genetic resources. For example, because LSD virus is secreted in semen (Irons et al. 2005), a trade ban exists on the product, even though it is not clear if the virus can be transmitted through the semen. Such bans impose heavy losses on countries exporting bull semen.

International trade agreements may therefore influence a country's LSD control policy. The use of efficacious vaccines, such as the attenuated Neethling vaccine (Ben-Gera et al. 2015), is economically attractive as long as vaccination does not trigger a trade restriction. In Ethiopia, vaccination is a sufficiently cheap way to reduce losses, despite the available vaccine's low efficacy (Gari et al. 2011).

References

Agianniotaki EI, Tasioudi KE, Chaintoutis SC, Iliadou P, Mangana-Vougiouka O, Kirtzalidou A, Alexandropoulos T, Sachpatzidis A, Plevraki E, Dovas CI, Chondrokouki E (2017) Lumpy skin disease outbreaks in Greece during 2015-16, implementation of emergency immunization and genetic differentiation between field isolates and vaccine virus strains. Vet Microbiol 201:78–84

Alemayehu G, Zewde G, Admassu B (2013) Risk assessments of lumpy skin diseases in Borena bull market chain and its implication for livelihoods and international trade. Trop Anim Health Prod 45:1153–1159

Ben-Gera J, Klement E, Khinich E, Stram Y, Shpigel NY (2015) Comparison of the efficacy of Neethling lumpy skin disease virus and x10RM65 sheep-pox live attenuated vaccines for the prevention of lumpy skin disease: the results of a randomized controlled field study. Vaccine 33:4837–4842

EFSA AHAW Panel (EFSA Panel on Animal Health and Welfare) (2015) Scientific opinion on lumpy skin disease. EFSA J 13(1):3986. https://doi.org/10.2903/j.efsa.2015.3986

Gari G, Bonnet P, Roger F, Waret-Szkuta A (2011) Epidemiological aspects and financial impact of lumpy skin disease in Ethiopia. Prev Vet Med 102:274–283

Irons PC, Tuppurainen ES, Venter EH (2005) Excretion of lumpy skin disease virus in bull semen. Theriogenology 63:1290–1297

Peck D, Bruce M (2017) The economic efficiency and equity of government policies on brucellosis: comparative insights from Albania and the United States of America. Rev Sci Tech Off Int Epiz 36:291–302

Sevik M, Dogan M (2017) Epidemiological and molecular studies on lumpy skin disease outbreaks in Turkey during 2014-2015. Transbound Emerg Dis 64(4):1268–1279

Thrusfield M (2005) The economics of infectious diseases. In: Veterinary epidemiology. Blackwell Science, London, pp 357–367

Tuppurainen ES, Oura CA (2011) Lumpy skin disease: an emerging threat to Europe, the Middle East and Asia. Transbound Emerg Dis 59:40–48

Geographic Distribution of Lumpy Skin Disease

4

Eeva S. M. Tuppurainen

A cattle disease causing skin nodules and fever was known to circulate in the Central Africa for years (Thomas and Mare 1945) before it was reported in Zambia in 1929 (Thomas 1945; MacOwan 1959). After sweeping throughout and causing substantial economic losses for the cattle farming in the Southern African region in the 1940s, the disease continued to spread towards the north, and currently it is present practically everywhere in Africa, including Madagascar (World Animal Health Information database, OIE WAHID Interface). Only Libya, Algeria, Morocco and Tunisia are still officially free of lumpy skin disease (LSD).

Many Middle Eastern countries import live cattle from the African Horn region (Shimshony and Economides 2006) where LSD is endemic. The first incursion of the disease in Egypt was reported in May 1988 followed by outbreaks in Israel in August 1989 (Yeruham et al. 1995). In 2006 LSD reoccurred in Egypt after being introduced by imported cattle from the Horn of Africa (El-Kholy et al. 2008). Outbreaks in the southern part of Israel swiftly followed (Brenner et al. 2009). In Oman LSD outbreaks occurred to such an extent in 2009 that it is currently considered endemic (Somasundaram 2011; Tageldin et al. 2014).

Between 2012 and 2013, new LSD outbreaks were reported in Israel (Ben-Gera et al. 2015). This time the disease was initially detected in beef cattle herds close to the northern borders with Lebanon and Syria. Swiftly more outbreaks started to appear also in dairy herds at the central part of Israel. The Israeli veterinary authorities responded by carrying out mass vaccinations, using a live attenuated LSDV vaccine together with Yugoslavian sheeppox virus (SPPV) RM65 strain containing vaccine, which successfully halt the spread of the disease.

Despite the widespread of LSD across the Middle East, no outbreaks have been reported by the Syrian veterinary authorities which are likely to be due to the civil war and disrupted veterinary infrastructure in the country.

First LSD outbreaks in Lebanon in 2012–2013 (ProMed 20130118.1505118) were controlled by immunizing the whole cattle population using RM65 vaccine. However, in 2016 new outbreaks were detected in the northern part of the country

© Springer International Publishing AG, part of Springer Nature 2018   11
E. S. M. Tuppurainen et al., *Lumpy Skin Disease*,
https://doi.org/10.1007/978-3-319-92411-3_4

which were likely to be associated with unauthorized cattle movements from Syria (ProMed 20160724.4365199).

Egypt also reported LSD outbreaks in 2012–2014, and currently the disease is considered as endemic. Jordan (Abutarbush et al. 2013) and Palestinian Autonomous Territories (ProMed 20130311.1581763) joined the list of infected countries in 2013. In Kuwait LSD was reported in late 2014 and early 2015 (ProMed 20150206.3147602) and in Saudi Arabia in spring 2015 (ProMed 20150430.3333997).

Subsequently in 2013, LSD spread to Iraq (Al-Salihi and Hassan 2015), and in 2014 outbreaks were also detected in the northwestern provinces of Iran, sharing borders with Iraq, Turkey, Azerbaijan and Armenia (ProMed 20140623.2561202) (Sameea Yousefi et al. 2016). Also both Saudi Arabia (ProMed 20150430.3333997) and Bahrain were affected in 2015.

Between 2013 and 2015, LSD swept across practically the whole territory of Turkey (ProMed 20130831.1915595) which is currently also considered as an endemic country (Timurkan et al. 2016). In Turkey the local attenuated SPP Bakirköy strain is used to immunize cattle against LSD.

In late 2014, LSD was reported in cattle farms located in the Karpas peninsula in the northern part of Cyprus (ProMed 20141205.3012426). The disease was effectively brought under control by conducting swift vaccination campaign comprising the whole cattle population in the region and using an attenuated LSDV containing vaccine.

In the Thrace region, the Evros river delta is a commonly used grazing land for free-ranging beef cattle on both sides of the border between Turkey and Greece. In August 2015, LSD outbreaks were first detected in cattle on the Turkish side but spread swiftly to herds on the Greek side (OIE Wahid, ProMed 20150821.3594203) (Tasioudi et al. 2016). A large-scale vaccination campaign was started, but due to delays in obtaining vaccines in sufficient numbers, the disease continued to spread to the central and western parts of Greece.

As anticipated, in 2016 the disease spread to Bulgaria (ProMed 20160415.4160177), Serbia (ProMed 20160609.4273545), Montenegro (ProMed 20160722.4360549), the former Yugoslav Republic of Macedonia (OIE WAHIS Interface), Kosovo (ProMed20160701.4321118) and Albania (ProMed 20160711.4337862).

From Turkey, Iraq and Iran, LSD spread to the northern Caucasus region, and incursions of LSDV were detected in Azerbaijan in 2014 (ProMed 20140719.2621294) (Zeynalova et al. 2016). First LSD cases in Armenia were detected in late December 2015, and the laboratory confirmation was obtained in early 2016 (ProMed 20160119.3948943). Georgia reported two LSD outbreaks in November 2016 (ProMed 20151019.3726848), and Kazakhstan was affected in late 2016 (ProMed 20160727.4374008).

Outbreaks were detected in the Russian states of Dagestan, Chechnya, North Ossetia, Kalmykia, Karachay-Cherkessia, Adygea and Kabardino-Balkaria and Ingushetia. Also the regions of Stavropol, Astrakhan, Volgograd, Tambov, Rostov and Samara, Ryazan, Voronezh and Krasnodar were affected (ProMed 20150904.3622855 and 20150921.3659823).

In the northern Caucasus, fast-moving LSDV is increasing the risk in the central parts of the Russian Federation, Ukraine, Moldova, Turkmenistan, Afghanistan and Pakistan.

In affected Balkan countries, mass vaccination of the whole cattle population was successfully completed by the end of 2016 and during 2017. Greece and Bulgaria implemented a total stamping-out policy. The rest of the infected countries culled initially all cattle in the affected farms, but after initiation of the vaccination campaigns, only animals are showing clinical signs. To date, nearly 100% vaccination coverage has been obtained in Southeast Europe, using vaccines with demonstrated efficacy against LSD, and the spread of the disease have been brought under control.

References

Abutarbush SM, Ababneh MM, Al Zoubi IG et al (2013) Lumpy skin disease in Jordan: disease emergence, clinical signs, complications and preliminary-associated economic losses. Transbound Emerg Dis. https://doi.org/10.1111/tbed.12177

Al-Salihi KA, Hassan IQ (2015) Lumpy skin disease in Iraq: study of the disease emergence. Transbound Emerg Dis 62:457–462. https://doi.org/10.1111/tbed.12386

Ben-Gera J, Klement E, Khinich E et al (2015) Comparison of the efficacy of Neethling lumpy skin disease virus and x10RM65 sheep-pox live attenuated vaccines for the prevention of lumpy skin disease: the results of a randomized controlled field study. Vaccine. https://doi.org/10.1016/j.vaccine.2015.07.071

Brenner J, BellaicheM GE, Elad D, Oved Z, Haimovitz M, Wasserman A, Friedgut O, Stram Y, Bumbarov V, Yadin H (2009) Appearance of skin lesions in cattle populations vaccinated against lumpy skin disease: statutory challenge. Vaccine 27:1500–1503

El-Kholy AA, Soliman HMT, Abdelrahman KA (2008) Polymerase chain reaction for rapid diagnosis of a recent lumpy skin disease virus incursion to Egypt. Arab J Biotechnol 11:293–302

MacOwan RDS (1959) Observation on the epizootiology of lumpy skin disease during the first year of its occurrence in Kenya. Bull Epizoot Dis Africa 7:7–20

Sameea Yousefi P, Mardani K, Dalir-Naghadeh B, Jalilzadeh-Amin G (2016) Epidemiological study of lumpy skin disease outbreaks in North-western Iran. Transbound Emerg Dis. https://doi.org/10.1111/tbed.12565

Shimshony A, Economides P (2006) Disease prevention and preparedness for animal health emergencies in the Middle East. Rev Sci Tech Int Des Epizoot 25:253–269

Somasundaram MK (2011) An outbreak of lumpy skin disease in a Holstein dairy herd in Oman: a clinical report. Asian J Anim Vet Adv 6:851–859. https://doi.org/10.3923/ajava.2011.851.859

Tageldin MH, Wallace DB, Gerdes GH et al (2014) Lumpy skin disease of cattle: an emerging problem in the Sultanate of Oman. Trop Anim Health Prod 46:241–246. https://doi.org/10.1007/s11250-013-0483-3

Tasioudi KE, Antoniou SE, Iliadou P et al (2016) Emergence of lumpy skin disease in Greece, 2015. Transbound Emerg Dis. https://doi.org/10.1111/tbed.12497

Thomas AD (1945) Lumpy skin disease: a new disease of cattle in the union. Farming S Afr 79:1–6

Thomas AD, Mare CVE (1945) Knopvelsiekte. J South African Vet Med Assoc 16:36–43

Timurkan MÖ, Özkaraca M, Aydın H, Sağlam YS (2016) The detection and molecular characterization of lumpy skin disease virus, northeast Turkey. Int J Vet Sci 5:44–47

Yeruham I, Nir O, Braverman Y et al (1995) Spread of lumpy skin disease in Israeli dairy herds. Vet Rec 137:91–93

Zeynalova S, Asadov K, Guliyev F et al (2016) Epizootology and molecular diagnosis of lumpy skin disease among livestock in Azerbaijan. Front Microbiol 7:1022. https://doi.org/10.3389/fmicb.2016.01022

Current Legislation and Trade Recommendations

5

Eeva S. M. Tuppurainen

In case of lumpy skin disease (LSD) incursion, national legislation needs to be in place to provide the competent veterinary authorities with means to effectively control and eradicate LSD and to carry out surveillance programmes. National legislation forms the basis for a contingency plan that is used in an exceptional disease situation. A good contingency plan describes administrative and logistic organisation, states the legal powers of the veterinary authority at all administrative levels and describes the measures to be conducted in order to halt the spread the disease. The legislation also indicates the source of funding for the control and eradication measures, such as but not limited to the costs of the vaccination campaign, implementation of stamping-out and associated measures, disposal of the carcasses and compensation farmers receive of cattle culled due to LSD infection or those animals died of LSD. A well-prepared contingency plan must be regularly updated, and simulation exercises should be organized for the staff at appropriate strategic levels. The Food and Agriculture Organization of the United Nations (FAO) has published a contingency plan template designed for the lumpy skin disease. The template is available and free to download on FAO website (http://www.fao.org/fileadmin/user_upload/reu/europe/documents/LSD_template.pdf).

Due to its economic impact, LSD is classified on the list of notifiable diseases by the World Organization for Animal Health (OIE). Standards and recommendations for safe international trade of live cattle and cattle products are set out in the Infection with Lumpy skin disease virus Chapter of the Terrestrial Animal Health Code (OIE 2016a). Terrestrial Code is an annually published reference document intended for use by veterinary authorities and other stakeholders involved in cattle import and export services. The measures recommended in the Terrestrial Code have to be formally adopted by the World Assembly of the Delegates of the OIE Member Countries. Recommendations are also set out for disease notification, provision of epidemiological information, control, surveillance and risk analysis.

Description on the diagnostic assays for the detection of LSD antigen and antibody as well as vaccines against LSD can be found in the OIE's Manual of

© Springer International Publishing AG, part of Springer Nature 2018
E. S. M. Tuppurainen et al., *Lumpy Skin Disease*,
https://doi.org/10.1007/978-3-319-92411-3_5

15

Diagnostic Tests and Vaccines for Terrestrial Animals (Terrestrial Manual) (OIE 2016b). The updated Lumpy skin disease Chapter was adopted in May 2016 and can be accessed via the OIE web site (http://www.oie.int/fileadmin/Home/eng/Health_ standards/tahm/2.04.13_LSD.pdf).

FAO provides practical advice for the effective disease control in the field. Detailed instructions are given for cleaning and disinfection of personnel, premises, vehicles, machinery, dairy equipment, milk storage tanks, other electric equipment as well as a captive bolt and firearms (FAO 2001). Safe disposal of contaminated feed, beddings, manure and effluent is also described.

In 2017 FAO published a field manual for lumpy skin disease intended for veterinarians, para-veterinarians, farmers and other stakeholders, providing detailed guidance for practical measures to be taken in case a suspected LSD case is detected in a holding (Tuppurainen et al. 2017).

Within the European Union (EU), several legislative acts apply to LSD, covering notification of the disease (82/894/EEC of 21 December 1982), intra-community trade in live animals and their products (90/425/EEC of 26 June 1990), as well as control and eradication measures applied in case of an outbreak within the EU (92/119/EEC of 17 December 1992).

Within the EU territory, vaccination against LSD using a live attenuated vaccine is currently not authorised in cattle but is allowed in an emergency situation if it doesn't affect the interests of the other member states.

In 2015 European Commission implemented country-specific decisions, allowing the emergency vaccinations against LSD virus in cattle in confined regions of Greece and Bulgaria. In late 2016, after the disease had swept practically throughout the Western Balkan region, in order to obtain uniformity in dealing with LSD outbreaks across the affected EU member and non-member states, one set of control and eradication measures was developed, and the country-specific decisions were repealed. In established directives, countries/zones were classified as infected or free-with-vaccination countries or zones (Commission Implementing Decision (EU) 2016/2009 and (EU) 2016/2008). This classification allowed different trade conditions for live animals and their products and possibility for bilateral trade agreement between trading countries. The reduced trade impact of vaccination on export of live cattle and their products enhanced the sustainability of preventive vaccination against LSD. Measures related to those cattle products that were considered safe or contain a negligible risk of transmission (such as meat and milk) were lifted or refined. Pasteurisation of milk is required only if destined for animal feed. Conditions were laid for transit of live bovines through LSD affected areas as well as for trade of unprocessed animal by-products, semen, embryos, ova, skins and hides. The new directive decreased the hesitation of those countries that have common borders with affected countries, to start pre-emptive vaccinating without a major impact to trade.

References

FAO (2001) Manual on procedures for disease eradication by stamping out. In: FAO Animal Health Manual. http://www.fao.org/docrep/004/Y0660E/Y0660E04.htm. Accessed 4 Jan 2017

OIE (2016a) Lumpy skin disease. In: OIE Terrestrial Animal Health Code. http://web.oie.int/eng/normes/mcode/en_chapitre_1.11.12.htm#rubrique_dermatose_nodulaire_contagieuse. Accessed 27 Aug 2016

OIE (2016b) Lumpy skin disease. In: OIE Manual of Diagnostic Tests and Vaccines for Terrestrial Animals

Tuppurainen E, Alexandro T, Beltrán-Alcrudo D (2017) Lumpy skin disease field manual. A manual for veterinarians, FAO Animal. Food and Agriculture Organization of the United Nations (FAO), Rome

Part I

Lumpy Skin Disease Virus

Taxonomy

6

Shawn Babiuk

The term "pox" originated from the English word "pock" (pustule), referring to skin lesions caused by a virus. This term was used for the nomenclature of viruses into families that cause pox lesions. A characteristic of poxviruses is the clinical manifestation of a "pox" lesion in the skin of infected animals. The classification of viruses into a genus is determined by the International Committee on Taxonomy of Viruses, which take into account many criteria including the type of disease they cause, host organisms, morphology, nucleic acid type and mode of replication. Members of each genus can be identified using serology as they often cross-react with each other. This is illustrated with the capripoxviruses, lumpy skin disease virus along with sheeppox virus and goatpox virus having no serotypes. The *Poxviridae* family consists of two subfamilies: the *Chordopoxvirinae* which infect vertebrates and the *Entomopoxvirinae* which infect insects. In the *Chordopoxvirinae* family, there are ten assigned genera and one unassigned genera (Table 6.1) (International Committee on Taxonomy of Viruses).

Avipoxvirus include the prototype fowlpox virus as well as several other members such as canarypox, pigeon pox, sparrowpox and other members which infect aves (birds). *Capripoxvirus* has three virus members which include sheeppox, goatpox and lumpy skin disease viruses which infect sheep, goats and cattle, respectively. The nomenclature is confusing as *capri* is derived from Latin for goat and caprinae refers to small ruminants like sheep and goats even though lumpy skin disease primarily infects cattle. *Cervidpoxvirus* has one member, the mule deerpox virus. *Crocodylidpoxvirus* has one member, the Nile crocodilepox virus. *Leporipoxvirus* has four members, hare fibroma virus, myxoma virus, rabbit fibroma virus and Squirrel fibroma virus. *Molluscipoxvirus* has one member, *Molluscum contagiosum virus*, a disease which infects humans. *Orthopoxvirus* have several members including the *Variola virus* which causes smallpox, vaccinia virus which was used as a vaccine to eradicate smallpox and monkeypox virus which can cause severe disease in monkeys and is zoonotic. Other members of the *Orthopoxvirus* genus are ectromelia virus which causes severe disease in mice, volepox virus, raccoonpox

© Springer International Publishing AG, part of Springer Nature 2018
E. S. M. Tuppurainen et al., *Lumpy Skin Disease*,
https://doi.org/10.1007/978-3-319-92411-3_6

Table 6.1 Classification of *Chordopoxvirinae* into genus and virus members

Genus	Virus members
Avipoxvirus	*Canarypox virus*
	Fowlpox[a]
	Juncopox virus
	Mynahpox virus
	Pigeon pox virus
	Psittacinepox virus
	Quailpox virus
	Sparrowpox virus
	Starlingpox virus
	Turkeypox virus
Capripoxvirus	Goatpox virus
	Lumpy skin disease virus
	Sheeppox virus[a]
Cervidpoxvirus	Mule deerpox virus[a]
Crocodylidpoxvirus	Nile crocodilepox virus[a]
Leporipoxvirus	Hare fibroma virus
	Myxoma virus[a]
	Rabbit fibroma virus
	Squirrel fibroma virus
Molluscipoxvirus	Molluscum contagiosum virus[a]
Orthopoxvirus	Camelpox virus
	Cowpox virus
	Ectromelia virus
	Monkeypox virus
	Raccoonpox virus
	Skunkpox virus
	Taterapox virus
	Vaccinia virus[a]
	Variola virus
	Volepox virus
Parapoxvirus	Bovine papular stomatitis virus
	Orf virus[a]
	Parapoxvirus of red deer in New Zealand
	Pseudocowpox virus
Suipoxvirus	Swinepox virus[a]
Unassigned	*Squirrelpox virus*
Yatapoxvirus	Tanapox virus
	Yaba monkey tumour virus[a]

International Committee on Taxonomy of Viruses
[a]Prototype virus

virus, skunkpox virus and *Camelpox virus* which causes severe disease in camels, leading to economic losses. Members of the parapoxviruses are bovine papular stomatitis virus, orf virus, parapoxvirus of red deer in New Zealand and pseudocowpox virus. These parapoxviruses cause skin lesions which are generally local self-limiting infections which resolve, although they can sometimes cause internal and external lesions which cause severe disease. In addition, some parapoxviruses such as orf have a broader host tropism and can cause skin lesions in humans. Swinepox virus is a member of the *Suipoxvirus* genus which infects pigs. *Squirrelpox virus* is currently unassigned to a genus. The *Yatapoxvirus* genus has tanapox virus and yaba monkey tumour virus as members.

Poxviruses are double-stranded DNA viruses in which the ends of the genome consist of terminal hairpin loops. The genome size of poxviruses is variable between viruses. Parapoxviruses genomes can be as small as 130 Kbp, whereas avipoxviruses genomes can be as large as 300 Kbp in size. Capripoxviruses have genomes around 150 Kbp in size. The number of coding genes for poxviruses is variable between different poxviruses.

The host range of poxviruses can vary greatly, with some poxviruses being extremely narrow, with others having a wide host range. For instance, vaccinia virus can infect a wide variety of animal species ranging from rodents to humans. In contrast, capripoxviruses have a narrow host range where sheeppox, goatpox and lumpy skin disease preferentially affect sheep goats and cattle, respectively (Babiuk et al. 2008). There are instances where a goatpox virus has been demonstrated to infect sheep to a lesser extent (Babiuk et al. 2009). It has also been demonstrated that the Kenyan sheep and goatpox vaccines derived from sheep are actually lumpy skin disease (Tuppurainen et al. 2014). The severity of disease caused by different poxviruses in their preferred host species also varies dramatically, causing clinical disease ranging from a devastating systemic disease leading to death to a local self-limiting infection which resolves without serious consequences.

References

Babiuk S, Bowden TR, Parkyn G, Dalman B, Manning L, Neufeld J, Embury-Hyatt C, Copps J, Boyle DB (2008) Quantification of lumpy skin disease virus following experimental infection in cattle. Transbound Emerg Dis 55:299–307

Babiuk S, Bowden TR, Parkyn G, Dalman B, Hoa DM, Long NT, Vu PP, Bieu do X, Copps J, Boyle DB (2009) Yemen and Vietnam capripoxviruses demonstrate a distinct host preference for goats compared with sheep. J Gen Virol 90:105–114

Tuppurainen ES, Pearson CR, Bachanek-Bankowska K, Knowles NJ, Amareen S, Frost L, Henstock MR, Lamien CE, Diallo A, Mertens PP (2014) Characterization of sheep pox virus vaccine for cattle against lumpy skin disease virus. Antiviral Res 109:1–6

Morphology

7

Shawn Babiuk

Poxviruses have a distinct morphology of brick-shaped that can be identified using electron microscopy. Parapoxviruses have an ovoid structure, whereas the capripoxviruses and the orthopoxviruses have a brick shape. The average size of lumpy skin disease virus (LSDV) particle is 294 ± 20 nm in length and 262 ± 22 nm in width (Kitching and Smale 1986). Using transmission electron microscopy, capripoxviruses display a biconcave core consisting of the genome which is in a triple-folded coil or tube. The core and two lateral bodies are surrounded by the capsid (Fig. 7.1). The members of *Capripoxvirus* genus cannot be distinguished by their size (Kitching and Smale 1986) as well as other orthopoxviruses. Parapox and capripoxviruses can be differentiated using scanning electron microscopy since capripoxviruses have a round smooth appearance (Fig. 7.2), whereas parapoxviruses have a distinct geometric pattern (Fig. 7.3). There are two forms of poxvirus virions, the intracellular mature virion (IMV) and the extracellular enveloped virion (EEV). These two forms of the virus are infectious even though they have different envelopes. Like other poxviruses the IMV primarily infects surrounding cells, and the EEV can infect both the surrounding and distant cells (Fenner et al. 1987; Moss 2006). The prototype of lumpy skin disease virus is a Neethling strain which was first isolated in South Africa (Alexander et al. 1957).

7

© Springer International Publishing AG, part of Springer Nature 2018
E. S. M. Tuppurainen et al., *Lumpy Skin Disease*,
https://doi.org/10.1007/978-3-319-92411-3_7

25

Fig. 7.1 Transmission
electron micrograph of
capripoxvirus. Photo by Lynn
Burton (National Centre for
Foreign Animal Disease)

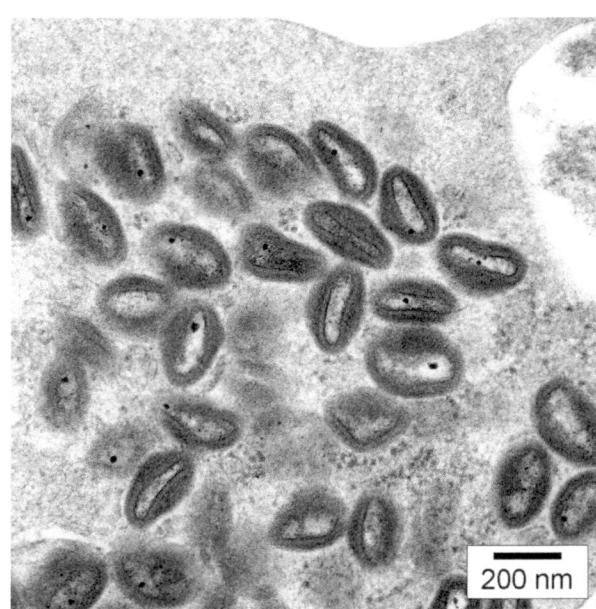

Fig. 7.2 Transmission
electron micrograph of
capripoxvirus. Photo by Lynn
Burton (National Centre for
Foreign Animal Disease)

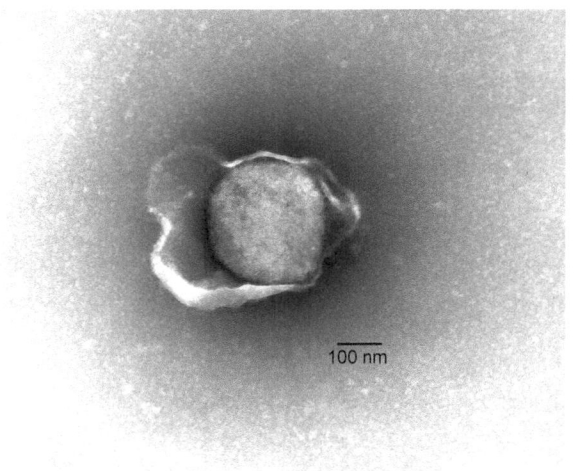

Fig. 7.3 Transmission electron micrograph of parapoxvirus. Photo by Lynn Burton (National Centre for Foreign Animal Disease)

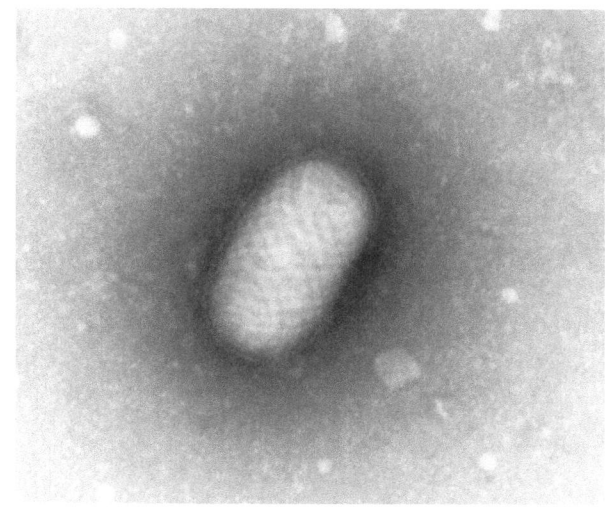

References

Alexander RA, Plowright W, Haig DA (1957) Cytopathogenic agents associated with lumpy skin disease of cattle. Bull Epizoot Dis Afr 5:489–492

Fenner F, Bachmann PA, Gibbs EPJ, Murphy FA, Studdert MJ, White DO (1987) Poxviridae. Veterinary virology. Academic Press, London

Kitching RP, Smale C (1986) Comparison of the external dimensions of capripoxvirus isolates. Res Vet Sci 41:425–427

Moss B (2006) Poxvirus entry and membrane fusion. Virology 344:48–54

Genome

Shawn Babiuk

Capripoxviruses are double-stranded DNA viruses that have hairpin loop ends which are similar to other poxviruses. Unfortunately, these regions have not yet been sequenced to directly compare these regions to the hairpin loop ends of other poxviruses. The genomes of members of the *Capripoxvirus* genus are approximately 150 Kbp in length. *Capripoxviruses* are similar to other poxviruses, with the genome being complex and encoding many genes. Before full-genome sequencing was available, restriction fragment analysis was used to compare and classify the relationship between capripoxviruses (Black et al. 1986; Gershon and Black 1988) and to some orthopoxvirus members (Gershon et al. 1989). These studies demonstrated the relationship between capripoxviruses and other poxviruses. The full-genome sequencing of lumpy skin disease virus together with bioinformatics revealed that lumpy skin disease virus encodes 156 putative genes (Tulman et al. 2001).

Lumpy skin disease contains many genes which are similar to other poxviruses which have been studied in greater detail (Table 8.1). This has allowed for the characterization of many of the open reading frames into their potential functions. The central region of lumpy skin disease virus between open reading frames ORF024 and ORF123 contains the homologues of poxvirus genes involved in the replication of the virus. These genes include viral DNA replication, viral transcription, RNA modification, viral assembly and viral structural proteins involved in both the IMV and EEV (Moss 2001). Surrounding the central region are genes that are involved in the host range, immune modulation and virulence. Like other poxviruses, lumpy skin disease virus has early, intermediate and late genes driven by specific promoters.

Lumpy skin disease virus has seven genes which are homologues of poxvirus genes involved in DNA replication such as LSDV039 DNA polymerase, LSDV077 DNA topoisomerase, LSDV082 uracil DNA glycosylase, LSDV083 NTPase, LSDV112 DNA polymerase processivity factor, LSDV133 DNA ligase and LSDV139 Ser/Thr protein kinase. There are 26 genes involved in the basic replication including transcription factors, mRNA transcription initiation, elongation

© Springer International Publishing AG, part of Springer Nature 2018
E. S. M. Tuppurainen et al., *Lumpy Skin Disease*,
https://doi.org/10.1007/978-3-319-92411-3_8

Table 8.1 Lumpy skin disease virus open reading frames and predicted function

ORF	Position (codon length)	Predicted structure/function	Promoter type
LSDV001	713–237 (159)	A52R-like family protein	
LSDV002	1179–787 (131)		
LSDV003	2151–1432 (240)	ER-localized apoptosis regulator	Early
LSDV004	2394–2224 (57)		
LSDV005	2446–2955 (170)	IL-10	
LSDV006	3664–2972 (231)	IL-1 receptor	
LSDV007	4753–3689 (355)	IFN-γ	
LSDV008	5664–4840 (275)	Family receptor	Late
LSDV009	6389–5700 (230)	α-Amanitin-sensitive protein. A52R-like protein	Early
LSDV010	6929–6444 (162)	LAP/PHD finger protein	Early
LSDV011	8118–6976 (381)	G protein-coupled CC chemokine receptor	Early
LSDV012	8860–8228 (211)	Ankyrin repeat protein	Early
LSDV013	9924–8902 (341)	IL-1 receptor	Early
LSDV014	10253–9987 (89)	IF2α-like PKR inhibitor	
LSDV015	10725–10243 (161)	IL-18 binding protein	
LSDV016	11031–10765 (89)	EGF-like growth factor	
LSDV017	11552–11025 (176)	Integral membrane protein, apoptosis regulator	Early
LSDV018	12034–11597 (146)	dUTPase	Early
LSDV019	13790–12084 (569)	Kelch-like protein	
LSDV020	14820–13858 (321)	Ribonucleotide reductase, small subunit	Early
LSDV021	15121–14864 (86)		Early/late
LSDV022	15500–15165 (112)		Early
LSDV023	15949–15734 (72)		Early
LSDV024	16676–16029 (216)		Late
LSDV025	17997–16657 (447)	Ser/Thr protein kinase, virus assembly	Late
LSDV026	18941–18036 (302)		
LSDV027	20866–18953 (638)	EEV maturation	
LSDV028	21985–20876 (370)	Palmitylated EEV envelope protein	Late
LSDV029	22624–22190 (145)		Early
LSDV030	23360–22704 (219)		Early
LSDV031	23434–23745 (104)	DNA-binding virion core phosphoprotein	Late
LSDV032	25176–23755 (474)	Poly(A) polymerase PAP$_L$	Early
LSDV033	27380–25176 (735)		
LSDV034	27925–27395 (177)	PKR inhibitor, host range	Early
LSDV035	28590–29795 (402)		
LSDV036	28591–27989 (201)	RNA polymerase subunit RPO30	Early
LSDV037	29807–31504 (566)		

(continued)

Table 8.1 (continued)

ORF	Position (codon length)	Predicted structure/function	Promoter type
LSDV038	31514–32311 (266)		
LSDV039	35343–32314 (1010)	DNA polymerase	
LSDV040	36053–35664 (130)	Potential redox protein, virus assembly	
LSDV041	36053–35664 (130)	Virion core protein	Late
LSDV042	38094–36043 (684)		
LSDV043	39144–38203 (314)	DNA-binding virion core protein, virus assembly	Late
LSDV044	39369–39154 (72)		Late
LSDV045	40200–39373 (276)	DNA-binding phosphoprotein	Early
LSDV046	40482–40249 (78)	IMV membrane protein	Late
LSDV047	41684–40503 (394)		
LSDV048	42978–41680 (433)	Virion core protein	Late
LSDV049	42984–45011 (676)	NPH-II, RNA helicase	
LSDV050	46801–45014 (596)	Metalloprotease, virion morphogenesis	Late
LSDV051	47124–47789 (222)	Putative transcriptional elongation factor	
LSDV052	47130–46801 (110)		Late
LSDV053	48136–47759 (126)	Glutaredoxin 2, virion morphogenesis	Late
LSDV054	48139–49449 (437)		
LSDV055	49453–49641 (63)	RNA polymerase subunit RPO7	Early/late
LSDV056	49644–50165 (174)		
LSDV057	51303–50185 (373)	Virion core protein	
LSDV058	51333–52112 (260)	Late transcription factor VLTF-1	Intermediate
LSDV059	52142–53149 (336)	Myristoylated protein	
LSDV060	53153–53887 (245)	Myristoylated IMV envelope protein	Late
LSDV061	53928–54203 (92)		Early
LSDV062	55172–54219 (318)		Late
LSDV063	55197–55955 (253)	DNA-binding virion core protein VP8	Late
LSDV064	55974–56366 (131)		Late
LSDV065	56326–56766 (147)		Late
LSDV066	56797–57327 (177)	Thymidine kinase	
LSDV067	57402–57995 (198)	Host range protein	Early
LSDV068	58056–59054 (333)	Poly(A) polymerase PAP$_S$	
LSDV069	58972–59526 (185)	RNA polymerase subunit RPO22	
LSDV070	59936–59538 (133)		
LSDV071	60022–63876 (1285)	RNA polymerase subunit RPO147	Early
LSDV072	64399–63887 (171)	Protein-tyrosine phosphatase, virus assembly	Late

(continued)

Table 8.1 (continued)

ORF	Position (codon length)	Predicted structure/function	Promoter type
LSDV073	64415–64984 (190)		
LSDV074	65952–64987 (322)	IMV envelope protein p35	
LSDV075	68378–65985 (798)	RNA polymerase-associated protein RAP94	Late
LSDV076	68522–69190 (223)	Late transcription factor VLTF-4	Early
LSDV077	69235–70185 (317)	DNA topoisomerase	
LSDV078	70208–70648 (147)		Late
LSDV079	70682–73207 (842)	mRNA-capping enzyme, large subunit	Early
LSDV080	73639–73175 (155)	Virion protein	
LSDV081	73641–74375 (245)	Virion protein	
LSDV082	74375–75028 (218)	Uracil DNA glycosylase	
LSDV083	75074–77431 (786)	NTPase; DNA replication	
LSDV084	77431–79335 (635)	Early transcription factor VETF$_{a+}$	Late
LSDV085	79363–79851 (163)	RNA polymerase subunit RPO18	
LSDV086	79895–80533 (213)	*mut* T motif	Early
LSDV087	80536–81294 (253)	*mut* T motif; gene expression regulator	
LSDV088	83210–81306 (635)	NPH-I; transcription termination factor	
LSDV089	84100–83240 (287)	mRNA-capping enzyme, small subunit; VITF	Early/late
LSDV090	85789–84143 (549)	Rifampin resistance protein, IMV assembly	
LSDV091	86268–85819 (150)	Late transcription factor VLTF-2	Intermediate/late
LSDV092	86996–86301 (232)	Late transcription factor VLTF-3	Intermediate
LSDV093	87220–86996 (75)		Late
LSDV094	89214–87232 (661)	Virion core protein P4b	Late
LSDV095	89824–89342 (161)	Virion core protein, virion morphogenesis	
LSDV096	89865–90374 (170)	RNA polymerase subunit RPO19	
LSDV097	91501–90377 (375)		Late
LSDV098	93666–91525 (714)	Early transcription factor VETF$_L$	Late
LSDV099	93723–94592 (290)	Intermediate transcription factor VITF-3	Early
LSDV100	94855–94622 (78)	IMV membrane protein	Late
LSDV101	97570–94859 (904)	Virion core protein P4a	Late
LSDV102	97585–98535 (317)		Late
LSDV103	99107–98538 (190)	Virion core protein	
LSDV104	99375–99175 (67)	IMV membrane protein	Late
LSDV105	99744–99460 (95)	IMV membrane protein	Late
LSDV106	99922–99764 (53)	Virulence factor	Early/late

(continued)

Table 8.1 (continued)

ORF	Position (codon length)	Predicted structure/function	Promoter type
LSDV107	100199–99915 (95)		Early/late
LSDV108	101316–100186 (377)	Myristoylated protein	Late
LSDV109	101922–101335 (196)	Phosphorylated IMV membrane protein	Late
LSDV110	101937–103376 (480)	DNA helicase; transcriptional elongation	
LSDV111	103584–103363 (74)		Late
LSDV112	103931–105220 (430)	DNA polymerase processivity factor	Early
LSDV113	103932–103588 (115)		
LSDV114	105192–105695 (168)		
LSDV115	105723–106877 (385)	Intermediate transcription factor VITF-3	Early
LSDV116	106911–110378 (1156)	RNA polymerase subunit RPO132	Early
LSDV117	110841–110398 (148)	Fusion protein, virus assembly	Late
LSDV118	111264–110845 (140)		Late
LSDV119	112173–111268 (302)		Late
LSDV120	112366–112145 (74)		Late
LSDV121	113309–112548 (254)	DNA packaging, virus assembly	
LSDV122	113441–114028 (196)	EEV glycoprotein	
LSDV123	114061–114573 (171)	EEV protein	Late
LSDV124	114604–115176 (191)		
LSDV125	115216–116079 (288)		Early
LSDV126	116141–116683 (181)	EEV glycoprotein	
LSDV127	116697–117515 (273)		
LSDV128	118424–117525 (300)	CD47-like protein	
LSDV129	118522–118890 (123)		
LSDV130	118962–119204 (81)		
LSDV131	119263–119745 (161)	Superoxide dismutase-like protein	Late
LSDV132	119783–120310 (176)		
LSDV133	120343–122019 (559)	DNA ligase	
LSDV134	122176–128250 (2025)	VAR B22R homologue	
LSDV135	128323–129402 (360)	IFN-α/β-binding protein	
LSDV136	129453–129911 (153)	A52R-like family protein	Early
LSDV137	129980–130984 (335)		Early/late
LSDV138	131017–131574 (186)	Ig domain, OX-2-like protein	
LSDV139	131616–132530 (305)	Ser/Thr protein kinase, DNA replication	
LSDV140	132565–133284 (240)	N1R/p28-like host range RING finger protein	
LSDV141	133336–134010 (225)	EEV host range protein	
LSDV142	134015–134416 (134)	Secreted virulence factor	
LSDV143	134456–135361 (302))	Tyrosine protein kinase-like protein	

(continued)

Table 8.1 (continued)

ORF	Position (codon length)	Predicted structure/function	Promoter type
LSDV144	135533–137173 (547)	Kelch-like protein	
LSDV145	137222–139123 (634)	Ankyrin repeat protein	
LSDV146	139255–140493 (413)	Phospholipase D-like protein	
LSDV147	140557–142050 (498)	Ankyrin repeat protein	
LSDV148	142101–143441 (447)	Ankyrin repeat protein	Early
LSDV149	143465–144475 (337)	Serpin	Early/late
LSDV150	144517–144999 (161)	A52R-like family protein	Early
LSDV151	145045–146694 (550)	Kelch-like protein	Early
LSDV152	146764–148230 (489)	Ankyrin repeat protein	Early
LSDV153	148278–148550 (91)		
LSDV154	148623–149342 (240)	ER-localized apoptosis regulator	Early
LSDV155	149595–149987 (131)		
LSDV156	150061–150537 (159)	A52R-like family protein	

Adapted from Tulman et al. (2001)

factors, termination factors, posttranslational modification of viral mRNA and RNA polymerase subunits. In addition there are 30 homologues of poxvirus proteins which are either involved in virion production and assembly or the structural proteins of the virus. These include viral core proteins LSDV048, LSDV057, LSDV063, LSDV094, LSDV095, LSDV126 and LSDV141; IMV membrane proteins LSDV074, LSDV100, LSDV104 and LSV105; and EEV membrane proteins LSDV028, LSDV122, LSDV123, LSDV126 and LSDV141.

Lumpy skin disease virus contains proteins likely involved in nucleotide metabolism including LSDV018 dUTPase, LSDV020 ribonucleotide reductase and LSDV066 thymidine kinase. In addition, lumpy skin disease virus contains cellular enzymes LSDV131 superoxide dismutase, LSDV143 tyrosine protein kinase and LSDV146 phospholipase D-like protein.

Lumpy skin disease virus contains several genes likely involved in determining host range as well as tissue and cell tropism such as LSDV016 EGF-like growth factor, LSDV034 PKR inhibitor, LSDV067 host range protein, LSDV104 N1R-/p28-like host range RING finger protein and ankyrin repeat proteins LSDV012, LSDV145, LSDV147, LSDV 148 and LSDV152.

Similar to other poxviruses, lumpy skin disease virus has several genes that are likely involved in the modulation or evasion of the host immune response. These include the LSDV005 IL-10; LSDV006 IL-1 receptor; LSDV007 IFN-γ; LSDV 10 LAP/PHD finger protein; LSDV011 G protein-coupled CC chemokine receptor; LSDV013 IL-1 receptor; LSDV014 and LSDV034 IFNα-like PKR inhibitor; LSDV015 IL-18 binding protein; Kelch-like proteins LSDV019, LSDV144 and LSDV151; LSDV034 IFNα-like PKR inhibitor; LSDV106 virulence factor; LSDV135 INF-α-/β-binding protein; LSDV142 secreted virulence factor; A52R-like family proteins LSDV001, LSDV009, LSDV136, LSDV150 and LSDV156; and serine proteinase inhibitors (serpins) LSDV149.

Sequencing of sheeppox and goatpox revealed that the genomes are very similar between members of the *Capripoxvirus* genus with 97% homology at the genetic level (Tulman et al. 2002). This study identified 156 genes are present in lumpy skin disease, with nine of these genes being nonfunctional in sheeppox and goatpox (Tulman et al. 2002). These genes are LSDV002, LSDV004, LSDV009, LSDV013, LSDV026, LSDV132, LSDV136, LSDV153 and LSDV155. These nine functional genes found in lumpy skin disease virus are likely to be involved in replication of lumpy skin disease virus in cattle. However, this has not yet been demonstrated by placing these functional genes in a sheeppox virus and evaluating if this virus would infect cattle.

Two field isolates of lumpy skin disease and the South African Onderstepoort vaccine (OBP) have been sequenced (Kara et al. 2003). This study revealed that 114 of the 156 genes had amino acid differences between the virulent Warmbath field isolate and the attenuated vaccine. These changes occurred throughout the genome, and the mechanism of attenuation could be due to either some mutations in a few critical genes or the combined effect of multiple changes. The Lumpyvax® by MSD and Lumpy Skin Vaccine for Cattle® by Onderstepoort Biological Product (OBP) vaccines have also been recently sequenced (Mathijs et al. 2016). This study revealed that these vaccines differ from the Neethling vaccine LW1959 with an amino acid modification T/M in LSDV056 and V/A in LSDV116 as well as two single-nucleotide deletions which do not affect the coding sequence. Lumpyvax has an amino acid modification G/V in LSDV037. The OBP vaccine has an 18-nt deletion in the terminal noncoding region as well as three single-nucleotide insertions and deletions affect the coding sequence (Mathijs et al. 2016).

References

Black DN, Hammond JM, Kitching RP (1986) Genomic relationship between capripoxviruses. Virus Res 5:277–292

Gershon PD, Black DN (1988) A comparison of the genomes of capripoxvirus isolates of sheep, goats, and cattle. Virology 164:341–349

Gershon PD, Ansell DM, Black DN (1989) A comparison of the genome organization of capripoxvirus with that of the orthopoxviruses. J Virol 63:4703–4708

Kara PD, Afonso CL, Wallace DB, Kutish GF, Abolnik C, Lu Z, Vreede FT, Taljaard LC, Zsak A, Viljoen GJ, Rock DL (2003) Comparative sequence analysis of the South African vaccine strain and two virulent field isolates of lumpy skin disease virus. Arch Virol 148:1335–1356

Mathijs E, Vandenbussche F, Haegeman A, King A, Nthangeni B, Potgieter C, Maartens L, Van Borm S, De Clercq K (2016) Complete genome sequences of the Neethling-like lumpy skin disease virus strains obtained directly from three commercial live attenuated vaccines. Genome Announc 4:e01255-16

Moss B (2001) *Poxviridae*: the viruses and their replication. In: Fields BN, Knipe DM, Howley PM, Chanock RM, Melnick JL, Monathy TP, Roizman B, Straus SE (eds) Fields virology, 4th edn. Lippincott, Williams and Wilkins, Philadelphia, PA, pp 2849–2883

Tulman ER, Afonso CL, Lu Z, Zsak L, Kutish GF, Rock DL (2001) Genome of lumpy skin disease virus. J Virol 75:7122–7130

Tulman ER, Afonso CL, Lu Z, Zsak L, Sur JH, Sandybaev NT, Kerembekova UZ, Zaitsev VL, Kutish GF, Rock DL (2002) The genomes of sheeppox and goatpox viruses. J Virol 76:6054–6061

Shawn Babiuk

Infection in cattle with lumpy skin disease virus (LSDV) can occur either by mechanical transmission of the virus by insect or tick vectors or experimentally by intravenous inoculation. The transmission of lumpy skin disease virus in the absence of vectors has not been experimentally demonstrated. This was performed by mixing infected cattle and naïve cattle. The results demonstrated that naïve cattle did not generate clinical signs of disease nor did they develop lumpy skin disease-specific antibodies (Carn and Kitching 1995). However, reported LSD outbreaks in winter-time without any vector activity indicate that it may happen in the field, and further studies using sufficient numbers of experimental animals need to be carried out to confirm this finding. It was demonstrated that *Aedes aegypti* female mosquitoes are capable of mechanical transmission of LSDV from infected to susceptible cattle. Mosquitoes that had fed upon lesions of LSDV-infected cattle were able to transmit virus to susceptible cattle over a period of 2–6 days post-infective feeding (Chihota et al. 2001). *Stomoxys calcitrans* has been demonstrated to be an efficient mechanical vector for sheeppox virus (Kitching and Mellor 1986) as well as goatpox virus (Mellor et al. 1987). Hard (ixodid) ticks *Rhipicephalus appendiculatus* (Tuppurainen et al. 2013a), *Rhipicephalus decoloratus* (Tuppurainen et al. 2013b) and *Amblyomma hebraeum* (Lubinga et al. 2015; Tuppurainen et al. 2011) have been demonstrated to mechanically transmit LSD. To date there is no evidence that LSDV can replicate in any insect or tick vector. In addition, infection can occur through the reuse of needles in vaccination programmes. Other modes of transmission for lumpy skin disease virus may be water troughs (Weiss 1968) and possibly through sexual transmission of semen containing lumpy skin disease virus (Annandale et al. 2014).

Following infection of cattle, the virus replicates in susceptible cells at levels undetectable using current molecular tests. Following experimental infection with *Capripoxvirus* in sheep and goats, primarily epithelial cells and macrophages have been demonstrated to contain *Capripoxvirus* by immunohistochemistry (Embury-Hyatt et al. 2012). With lumpy skin disease virus infection in cattle, immunohistochemistry revealed that several different cells contain lumpy skin disease viral

© Springer International Publishing AG, part of Springer Nature 2018   37
E. S. M. Tuppurainen et al., *Lumpy Skin Disease*,
https://doi.org/10.1007/978-3-319-92411-3_9

antigen, including keratinocytes, hair follicle epithelium, fibroblasts and interstitial macrophages, infiltrating the dermis, subcutis and parenchyma of lymph nodes (Babiuk et al. 2008; Awadin et al. 2011). After biting insects have transmitted the virus, replication continues in the blood and skin cells until viremia is detectable starting usually at 6 days post-infection. During viremia, both the mature and immature virus particles are disseminated throughout the body into susceptible tissues. Viremia lasts for about 9 days and sometimes longer, until antibody responses are elicited to neutralize the virus and stop viremia. Lumpy skin disease virus is shed at mucosal sites including nasal, oral and conjunctival secretions at least 1 week after viremia is cleared (Babiuk et al. 2008). It has been demonstrated that cattle can be viremic in the absence of clinical disease (Tuppurainen et al. 2005). The level of lumpy skin disease virus detected using virus isolation in the blood is at the limit of detection and is difficult to quantitate. In contrast high levels up to 10^5 plaque forming units (PFU)/mg of LSDV can be isolated from skin nodules and nasal turbinates (Babiuk et al. 2008). Lumpy skin disease virus can be excreted in semen, and live virus was isolated in experimentally infected bulls up to 22 days post-infection (Weiss 1968). In a more recent study, virus was isolated up to 42 days post-infection with the viral genome detected by PCR from 6 days post-inoculation to 159 days post-inoculation (Irons et al. 2005). Lumpy skin disease can also be secreted in the milk of infected cows (Weiss 1968; T. Alexandrov, personal communication).

Following the onset of viremia, around 7 days post-inoculation, skin nodules start to appear on the cattle as well as characteristic pox lesion on susceptible tissues which contain epithelial cells. These susceptible tissues include, mucous membranes of the mouth including the inside of the lips, gingivae and dental pads, tongue soft pallet, pharynx, epiglottis as well as the rumen, reticulum, omasum and abomasum. In addition, the mucous membranes of the nasal cavity including turbinate, trachea and lungs are susceptible. Depending on the severity of disease, the tissues which are affected will vary, with more tissues being infected with severe disease. The skin is the most commonly affected tissue where virus replication will occur. Some infected animals will have pox lesions predominantly in the skin with the absence of internal lesions, whereas some animals will have skin lesions as well as internal lesions in the lung and/or forestomach of the digestive tract. In addition, skin lesions contain the highest level of viral replication determined by both real-time PCR, up to 10^8 genome copies/µg as well as virus isolation 10^5 PFU/mg compared to other susceptible tissues. The characteristic lesions caused by poxviruses are due to the replication of the virus and immature virus particles infecting neighbouring cells causing the gross pathology of the lesions. There are high levels of virus found in the lesions, whereas tissue without any lesion does not generally contain high levels of virus. However, it was demonstrated in sheep and goats experimentally infected with sheeppox and goatpox that normal appearing skin may actually have high levels of virus replication (Bowden et al. 2008; Babiuk et al. 2009). It is possible that in cattle, severely infected with LSD, the skin that appears normal may have high levels of LSDV present. However, in experimentally infected cattle that did not have severe LSD, there was no LSDV isolated in the normal skin (Babiuk et al. 2008). It is likely that the high levels of virus found in the skin lesions along with virus in the blood are important for mechanical transmission to insect vectors or ticks when they feed on infected cattle.

Although LSDV is considered quite host specific for cattle, there have been observations of other animal species developing clinical disease. There is one documented case where LSDV has caused clinical disease in sheep. This occurred in Kenya in a flock of sheep that were mixed with cattle (Burdin and Prydie 1959). There have not been any other instances reported of LSDV causing disease in sheep or goats. However, due to the limited use of sequencing to determine the actual identity of the *Capripoxvirus*, it is possible that there have been other occurrences of LSDV infections in sheep and/or goats. Several different ruminant species may be susceptible to LSD under the right conditions. The giraffe and impala have been infected with LSDV by experimental inoculation of LSDV (Young et al. 1970). The Asian water buffalo (Ali et al. 1990) seems to be susceptible, and LSDV was detected in milk from water buffalo (Sharawi and Adb El-Rahim 2011). In an Arabian oryx (Greth et al. 1992), LSD was suspected, but it was not confirmed if the virus was indeed LSDV and not sheeppox or goatpox. In springbok samples, *Capripoxvirus* has been detected (Lamien et al. 2011). As lumpy skin disease spreads into new territory, it is possible that other ruminant species may be susceptible and develop clinical disease.

Animals affected by capripoxviruses will eventually clear the infection and do not become carriers. The length of time to clear capripoxvirus infection is longer compared to many respiratory virus infections as well as many vector transmitted viruses. This is due to the stability of *Capripoxvirus* in the skin and internal oral/ nasal lesions. Closely related sheeppox and goatpox viral, genome have been detected in oral and nasal swabs in sheep and goats for several weeks following infection (Bowden et al. 2008). Following experimental infection of LSD in cattle, the duration of detecting genome in oral and nasal swabs was much shorter compared to sheeppox and goatpox. It is difficult to compare these studies as the clinical disease was much more severe in the sheep and goats compared to the LSD in cattle. More studies are needed to evaluate the duration of nasal and oral shedding by real-time PCR in severely infected cattle. Lumpy skin disease viral genome was detected 42 days following infection in skin lesions in cattle (Babiuk et al. 2008).

References

Ali AA, Esmat M, Attia H, Selim A, Abdel-Hamid YM (1990) Clinical and pathological studies on lumpy skin disease in Egypt. Vet Rec 127:549–550

Annandale CH, Holm DE, Ebersohn K, Venter EH (2014) Seminal transmission of lumpy skin disease virus in heifers. Transbound Emerg Dis 61:443–448

Awadin W, Hussein H, Elseady Y, Babiuk S, Furuoka H (2011) Detection of lumpy skin disease virus antigen and genomic DNA in formalin-fixed paraffin-embedded tissues from an Egyptian outbreak in 2006. Transbound Emerg Dis 58:451–457

Babiuk S, Bowden TR, Parkyn G, Dalman B, Manning L, Neufeld J, Embury-Hyatt C, Copps J, Boyle DB (2008) Quantification of lumpy skin disease virus following experimental infection in cattle. Transbound Emerg Dis 55:299–307

Babiuk S, Bowden TR, Parkyn G, Dalman B, Hoa DM, Long NT, Vu PP, Bieu do X, Copps J, Boyle DB (2009) Yemen and Vietnam capripoxviruses demonstrate a distinct host preference for goats compared with sheep. J Gen Virol 90:105–114

Bowden TR, Babiuk SL, Parkyn GR, Copps JS, Boyle DB (2008) Capripoxvirus tissue tropism and shedding: a quantitative study in experimentally infected sheep and goats. Virology 371:380–393

Burdin ML, Prydie J (1959) Observations on the first outbreak of lumpy skin disease in Kenya. Bull Epizoot Dis Afr 7:21–26

Carn VM, Kitching RP (1995) An investigation of possible routes of transmission of lumpy skin disease virus (Neethling). Epidemiol Infect 114:219–226

Chihota CM, Rennie LF, Kitching RP, Mellor PS (2001) Mechanical transmission of lumpy skin disease virus by Aedes aegypti (Diptera: Culicidae). Epidemiol Infect 126:317–321

Embury-Hyatt C, Babiuk S, Manning L, Ganske S, Bowden TR, Boyle DB, Copps J (2012) Pathology and viral antigen distribution following experimental infection of sheep and goats with capripoxvirus. J Comp Pathol 146:106–115

Greth A, Gourreau JM, Vassart M, Nguyen-Ba-Vy WM, Lefevre PC (1992) Capripoxvirus disease in an Arabian oryx (Oryx leucoryx) from Saudi Arabia. J Wildl Dis 28:295–300

Irons PC, Tuppurainen ES, Venter EH (2005) Excretion of lumpy skin disease virus in bull semen. Theriogenology 63:1290–1297

Kitching RP, Mellor PS (1986) Insect transmission of capripoxvirus. Res Vet Sci 40:255–258

Lamien CE, Lelenta M, Goger W, Silber R, Tuppurainen E, Matijevic M, Luckins AG, Diallo A (2011) Real time PCR method for simultaneous detection, quantitation and differentiation of capripoxviruses. J Virol Methods 171:134–140

Lubinga JC, Tuppurainen ES, Mahlare R, Coetzer JA, Stoltsz WH, Venter EH (2015) Evidence of transstadial and mechanical transmission of lumpy skin disease virus by Amblyomma hebraeum ticks. Transbound Emerg Dis 62:174–182

Mellor PS, Kitching RP, Wilkinson PJ (1987) Mechanical transmission of capripox virus and African swine fever virus by Stomoxys calcitrans. Res Vet Sci 43:109–112

Sharawi SS, Abd El-Rahim IH (2011) The utility of polymerase chain reaction for diagnosis of lumpy skin disease in cattle and water buffaloes in Egypt. Rev Sci Tech 30:821–830

Tuppurainen ES, Venter EH, Coetzer JA (2005) The detection of lumpy skin disease virus in samples of experimentally infected cattle using different diagnostic techniques. Onderstepoort J Vet Res 72:153–164

Tuppurainen ES, Stoltsz WH, Troskie M, Wallace DB, Oura CA, Mellor PS, Coetzer JA, Venter EH (2011) A potential role for ixodid (hard) tick vectors in the transmission of lumpy skin disease virus in cattle. Transbound Emerg Dis 58:93–104

Tuppurainen ES, Lubinga JC, Stoltsz WH, Troskie M, Carpenter ST, Coetzer JA, Venter EH, Oura CA (2013a) Mechanical transmission of lumpy skin disease virus by Rhipicephalus appendiculatus male ticks. Epidemiol Infect 141:425–430

Tuppurainen ES, Lubinga JC, Stoltsz WH, Troskie M, Carpenter ST, Coetzer JA, Venter EH, Oura CA (2013b) Evidence of vertical transmission of lumpy skin disease virus in Rhipicephalus decoloratus ticks. Ticks Tick Borne Dis 4:329–333

Weiss KE (1968) Lumpy skin disease virus. Virol Monogr 3:111–113

Young E, Basson PA, Weiss KE (1970) Experimental infection of game animals with lumpy skin disease virus (prototype strain Neethling). Onderstepoort J Vet Res 37:79–87

Propagation of the Virus In Vitro

10

Shawn Babiuk

Capripoxviruses comprising sheeppox, goatpox and lumpy skin disease virus (LSDV) have a tropism for epithelial cells. This tropism allows capripoxviruses to be propagated in a wide variety of cell types from cattle, goat and sheep origin with virus titres up to a titre of 10^6 $TCID_{50}$ per ml. These cell types include cells from tissues, including the kidney, testes, adrenal, thyroid, skin and muscle. There are currently no cell culture systems which can differentiate lumpy skin disease from sheeppox and goatpox. The tropism and the cytopathic effect for all the evaluated cells are identical between capripoxviruses.

Following inoculation of LSDV to a monolayer of susceptible cells, the virus interacts with currently unknown viral glycoprotein cellular receptors and then enters the cell through fusion with the cell membrane (Moss 2006). The number of viral protein and host cell interactions is large due to the numerous capripoxvirus proteins. Following the viral/cell interaction, the fusion of the virus with the plasma membrane of the cell occurs, and the viral core is released into the cytoplasm of the cell. The lumpy skin disease virus genome then uncoats from the structural proteins in the cytoplasm, and the early genes start transcription, followed by the intermediate and late genes. Capripoxviruses have their own DNA polymerase and therefore replicate in the cytoplasm of the host cell. Intracellular mature particles are produced in the cell cytoplasm, and some of these virus particles leave the cell and incorporate the host cell plasma membrane forming extracellular enveloped virus. Infected cells induce the formation of distinct plaques at low virus concentrations (Soman and Singh 1980) with a cytopathic effect characterized by elongated cells (Jassim and Keshavamurthy 1981).

Infection of cells by *Capripoxvirus* can be determined by the presence of intracytoplasmic eosinophilic inclusion bodies (Weiss 1968; Prozesky and Barnard 1982), which can be detected using microscopic examination of a haematoxylin- and eosin-stained LSDV-infected monolayer of cells. The site of capripoxvirus virion assembly is in cytoplasmic inclusion bodies, type B. The egress mechanism for *Capripoxvirus* is either by budding from the cell through the Golgi membrane to

© Springer International Publishing AG, part of Springer Nature 2018
E. S. M. Tuppurainen et al., *Lumpy Skin Disease*,
https://doi.org/10.1007/978-3-319-92411-3_10

form extracellular enveloped virus (EEV) or through the rupture of host cell where intracellular mature virion (IMV) is released and can infect surrounding cells. The mature virions collect in type A inclusions.

The most commonly used cells for the propagation of capripoxviruses are primary lamb kidney or primary lamb testis cells (Ferris and Plowright 1958; Hess et al. 1963; Kalra and Sharma 1981; Plowright and Witcomb 1959; Zhou et al. 2004). Primary cells were traditionally used for *Capripoxvirus* propagation and require a fresh stock of cells isolated from animals in a labour-intensive and costly manner. In addition, primary cells do not lend themselves to standardization and quality control due to the heterogeneous cell population as well as genetic differences between the donor animals from which they are derived. The susceptibility to virus infection can also vary between different lots of primary cells. In addition, each batch of primary cells requires testing to ensure that the cells are free of adventitious agents such as retroviruses, which are expensive and time-consuming. In rare occasions other contaminating viruses such as bluetongue virus have been found vaccines used against LSDV (Bumbarov et al. 2016) likely due to primary cells being contaminated. Primary cells can display cell artefacts including unevenness of the primary lamb kidney (LK) cells in the monolayer making it challenging to determine endpoint titres because of difficulties in distinguishing virus-specific cytopathic effects. Despite these caveats, propagation of LSDV as well as sheeppox and goatpox vaccines and virus isolation are performed in primary cells.

Foetal bovine muscle cells (Binepal et al. 2001), MDBK cells and OA3.Ts cells have been used in laboratories for diagnostics (Babiuk et al. 2007). Vero cells have been demonstrated to be able to replicate capripoxviruses poorly compared to OA3. Ts cells and thus are not recommended for virus isolations (Babiuk et al. 2007). In addition, capripoxviruses can be grown in chick embryos (Van Den Ende et al. 1949) as well as on the chorioallantoic membrane of embryonated chicken eggs producing pocks on the membrane (Van Rooyen et al. 1969). Due to the lumpy skin disease virus being transmitted by insects and ticks, tick cell lines were evaluated for their ability to replicate lumpy skin disease virus. This study revealed that tick cell lines were not able to propagate lumpy skin disease virus (Tuppurainen et al. 2015).

Capripoxviruses are generally slow growing viruses where the observed cytopathic effect of virus replication can be seen approximately 3–4 days following infection. The cytopathic effect of LSDV forms plaques on OA3.Ts cells is characterized by bubbling of the smooth monolayer of OA3.Ts cells (Fig. 10.1). These plaques increase in size with time. When higher levels of virus are used to infect OA3.Ts cells, the monolayer will be destroyed, and cells will a have a spindle-like appearance (Fig. 10.1).

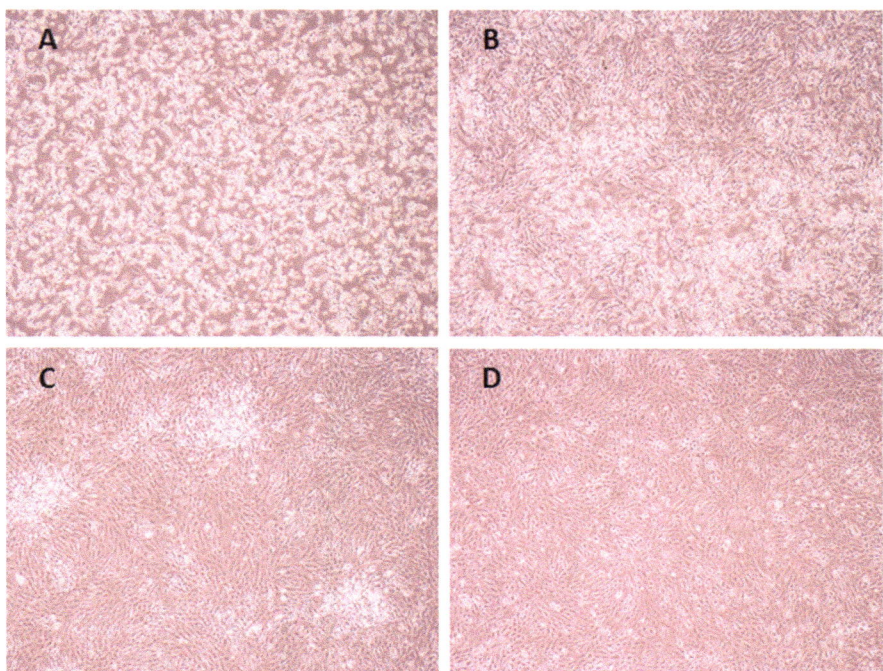

Fig. 10.1 Cytopathic effect of lumpy skin disease at different multiplicity of infections (MOI) 5 days following inoculation (**a**) MOI 0.01, (**b**) MOI 0.001, (**c**) MOI 0.0001, and (**d**) uninfected. Photos by Graham Blyth (National Centre for Foreign Animal Disease)

References

Babiuk S, Parkyn G, Copps J, Larence JE, Sabara MI, Bowden TR, Boyle DB, Kitching RP (2007) Evaluation of an ovine testis cell line (OA3.Ts) for propagation of capripoxvirus isolates and development of an immunostaining technique for viral plaque visualization. J Vet Diagn Invest 19:486–491

Binepal YS, Ongadi FA, Chepkwony JC (2001) Alternative cell lines for the propagation of lumpy skin disease virus. Onderstepoort J Vet Res 68:151–153

Bumbarov V, Golender N, Erster O, Khinich Y (2016) Detection and isolation of Bluetongue virus from commercial vaccine batches. Vaccine 34:3317–3323

Ferris RD, Plowright W (1958) Simplified methods for the production of monolayers of testis cells from domestic animals, and for serial examination of monolayer cultures. J Pathol Bacteriol 75:313–318

Hess WR, May HJ, Patty RE (1963) Serial cultures of lamb testicular cells and their use in virus studies. Am J Vet Res 24:59–63

Jassim FA, Keshavamurthy BS (1981) Cytopathic changes caused by sheep pox virus in secondary culture of lamb testes cells. Bull Off Int Epizoot 93:1401–1410

Kalra SK, Sharma VK (1981) Adaptation of Jaipur strain of sheeppox virus in primary lamb testicular cell culture. Indian J Exp Biol 19:165–169

Moss B (2006) Poxvirus entry and membrane fusion. Virology 344:48–54

Plowright W, Witcomb MA (1959) The growth in tissue cultures of a virus derived from lumpy-skin disease of cattle. J Pathol Bacteriol 78:397–407

Prozesky L, Barnard BJH (1982) A study of the pathology of lumpy skin disease in cattle. Onderstepoort J Vet Res 49(3):167–175

Soman JP, Singh IP (1980) Plaque formation by sheep pox virus adapted to lamb kidney cell culture. Indian J Exp Biol 18:313–314

Tuppurainen ES, Venter EH, Coetzer JA, Bell-Sakyi L (2015) Lumpy skin disease: attempted propagation in tick cell lines and presence of viral DNA in field ticks collected from naturally-infected cattle. Ticks Tick Borne Dis 6:134–140

Van Den Ende M, Don PA, Kipps A (1949) The isolation in eggs of a new filterable agent which may be the cause of bovine lumpy skin disease. J Gen Microbiol 3:174–183

Van Rooyen PJ, Kumm NAL, Weiss KE (1969) The optimal conditions for the multiplication of Neethling-type lumpy skin disease virus in embryonated eggs. Onderstepoort J Vet Res 36:165–174

Weiss KE (1968) Lumpy skin disease virus. Virol Monogr 3:111–113

Zhou JS, Ma HL, Guo QS (2004) Culturing of ovine testicular cells and observation of pathological changes of the cell inoculated with attenuated sheep pox virus. Chinese J Vet Sci Technol 34:71–74

Shawn Babiuk

Lumpy skin disease virus is a stable virus which can be inactivated by temperature at 55 °C for 2 h, 60 °C for 1 h or 65 °C for 30 min (OIE 2016). The virus is stable for extended periods of time for at least 10 years in skin lesions that are frozen at −80 °C. Freezing and thawing LSDV will reduce the virus titre slightly (Haig 1957). Lumpy skin disease virus is susceptible to high alkaline or acid pH, although it is stable between pH 6.6 and 8.6 (Weiss 1968). It can be inactivated by (20%) chloroform, (1%) formalin, and detergents such as sodium dodecyl sulphate and detergents containing lipid solvents. Also phenol 2% in 15 min, sodium hypochlorite 2–3%, iodine compounds (1:33) dilution, Virkon® (2%) and quaternary ammonium compounds (0.5%) can be used as disinfectant. Lumpy skin disease virus can be inactivated by ultraviolet light. Due to the sensitivity to sunlight, lumpy skin disease vaccines should be manufactured in dark glass bottles. Lumpy skin disease virus is viable for 35 days at 28 °C in the phosphate buffered saline (Tuppurainen et al. 2015).

Environmental contamination occurs since LSDV is stable for long periods at ambient temperature especially in dried scabs which can contaminate beddings. Due to the shedding of LSDV through oral and nasal secretions, animal holding facilities will also be contaminated. However, as LSD is mainly transmitted by biting arthropods, the importance of such contamination in virus spread is not clear. Stables can provide shade to protect *Capripoxvirus* from being inactivated by ultraviolet light.

In EFSA Scientific Opinion on LSD, the survival time of the virus in different matrixes is summarized based on the currently available literature (EFSA 2015).

References

EFSA AHAW Panel (EFSA Panel on Animal Health and Welfare) (2015) Scientific opinion on lumpy skin disease. EFSA J 13(1):3986, 73 pp. doi:https://doi.org/10.2903/j.efsa.2015.3986

Haig DA (1957) Lumpy skin disease. Bull Epizoot Dis Africa 5:421–430

© Springer International Publishing AG, part of Springer Nature 2018
E. S. M. Tuppurainen et al., *Lumpy Skin Disease*,
https://doi.org/10.1007/978-3-319-92411-3_11

OIE (World Organization for Animal Health) (2016) General recommendations on disinfection and disinfection. In: Terrestrial Animal Health Code, Paris

Tuppurainen ES, Venter EH, Coetzer JA, Bell-Sakyi L (2015) Lumpy skin disease: attempted propagation in tick cell lines and presence of viral DNA in field ticks collected from naturally-infected cattle. Ticks Tick Borne Dis 6:134–140

Weiss KE (1968) Lumpy skin disease virus. Virol Monogr 3:111–113

Immunity

12

Shawn Babiuk

There are many factors that can play a role in the susceptibility of infection to lumpy skin disease virus (LSDV). The interaction between the host's immune system and the virus determines the outcome. Following natural or experimental infection of cattle with virulent LSDV, virus neutralizing antibodies are elicited. These antibodies are detected starting around 15 days following infection, with the titres of the antibodies increasing over the next 2 weeks and then wane. With capripoxvirus infections, mild infections generate lower antibody responses compared to severe infections (Bowden et al. 2009). Unfortunately, following vaccination with several of the currently available vaccines, in many times only low or undetectable levels of virus neutralizing antibodies are elicited.

The role of antibodies in protection against CaPV was demonstrated by passive transfer of sera from infected sheep, which protected the recipient sheep against CaPV challenge (Kitching 1986), suggesting that antibodies alone are sufficient for protection. Unfortunately, passive transfer experiments have not been done in cattle. However, since sheeppox causes a more severe disease in sheep, it is very likely that passive transfer would work in cattle against lumpy skin disease (LSD). The use of passive transfer of human vaccinia-neutralizing antibodies to macaques was able to provide protection against monkeypox (Edghill-Smith et al. 2005), demonstrating that passive transfer can be effective for poxviruses. The role of antibody-dependent cytotoxicity was demonstrated using cells from humans vaccinated with vaccinia (Perrin et al. 1977). Antibody-dependent cell cytotoxicity has not been studied for capripoxvirus infections.

The role of passive immunity with LSDV has been described in unpublished observations by Westhuizen. Calves born to immunized cows were protected by antibodies derived from the colostrum up to 6 months (Weiss 1968). Since antibodies following vaccination with LSDV vaccines can be at levels beneath current assays detection limits, it is difficult to see how antibodies can be passed through colostrum if they are not present in the sera. For these reasons, further studies need to be performed evaluating the presence of antibodies against LSDV in the sera of the vaccinated cow, the colostrum as well as the sera from the calf over a 6-month period.

The role of different cells involved in the generation of immunity to LSDV has not been fully elucidated; however, the role of cellular immunity has been well characterized using other poxvirus infections. The role of CD8+ T-cell function was determined with ectromelia using several different knockout mice. IFN-γ or perforin gene knockout mice died early in infection, whereas B cell knockout mice and MHC class II knockout mice died later (Chaudhri et al. 2006). This study indicates how B-cell function becomes critical after the CD8+ T-cell effector phase of infection subsides. Additional studies in mice with ectromelia demonstrated that antibody is essential, but not CD4+ or CD8+ T cells to recover from a secondary poxvirus infection as primed mice that lack B cells, MHC class II or CD40 succumbed to secondary infection (Panchanathan et al. 2006). With monkeypox virus infection in macaques, it was demonstrated that antibody responses were essential, but not CD4+ or CD8+ T cells, using antibody-mediated depletion of these cells (Edghill-Smith et al. 2005). In humans, the complication of progressive vaccinia occurred at a rate of approximately one in a million vaccinations with the smallpox vaccine. The condition is due to defective cell-mediated immunity, illustrating the critical role that cell-mediated immunity plays in controlling the early stages of poxvirus infection (Hathaway et al. 1965; Bray and Wright 2003).

The role of natural killer (NK) cells was examined with both antibody depletion and NK-deficient knockout mice using ectromelia (Parker et al. 2007). The results of this study demonstrated that a lack of NK cells increased the susceptibility of resistant mice to lethal disease, indicating the importance of NK cells in controlling poxvirus infection with ectromelia (Parker et al. 2007).

There have been no studies in cattle that have used antibody-mediated depletion of B cell, CD4+ and/or CD8+ T cells as well as the generation of gene knockout cattle to determine the roles of these cells in LSDV infections. The knowledge gleaned using orthopoxviruses indicates that cellular immune responses are critical to respond to the early stages of infection and antibody responses required for clearance of the infection. Both humoral and cellular immune responses are as important in responding to capripoxvirus infections. Antibody-mediated protection was demonstrated through passive transfer of antibodies and cellular immunity by protection in vaccinated animals in the absence of detectable antibody responses. However, it is likely that both antibody and cellular responses are required for immunity to be generated against capripoxviruses. This is because in passive transfer experiment, the cellular component of the immune system is still intact and is being stimulated to generate immunity; likewise following vaccination it is likely that B cells are generating antibody responses that are under the limit of detection of the current assays used. The correlates of protection are not known for LSDV. This is due to vaccinated cattle being protected in the absence of detectable antibody responses, making antibody evaluation not a suitable correlate of protection and cellular assays to measure T-cell responses difficult to develop, use and validate a cellular correlate of protection.

Studies in humans vaccinated against smallpox have demonstrated that both B-cell (Crotty et al. 2003) and T-cell memory responses are long lived (Amara et al. 2004; Hammarlund et al. 2003). With capripoxviruses it is likely that lifelong immunity is generated in animals that have been infected and generate clinical disease. For animals that do not develop clinical disease or in animals that have been vaccinated, the duration of immunity may not last for the life of the animal. Performing similar studies in cattle to evaluate immune memory responses to LSDV could answer the questions regarding the duration of immunity against LSDV. The duration of immunity using the recombinant KS-1 rinderpest vaccine was assessed against LSDV using a challenge with LSDV and full protection in cattle for 2 years following vaccination, and partial protection after 3 years was observed (Ngichabe et al. 2002).

The capripoxvirus antigens involved in neutralizing antibody responses are not currently known. For the orthopoxvirus vaccinia, there are nine specific B-cell epitopes (Moutaftsi et al. 2010). Five of these proteins have been demonstrated to elicit protective neutralizing responses in mice H3 (Rodriguez et al. 1985), A27 (Gordon et al. 1991), L1 (Wolffe et al. 1995) B5R neutralizing (Galmiche et al. 1999) and D8 (Hsiao et al. 1999). In addition, A33 induces protection with non-neutralizing antibodies (Galmiche et al. 1999). A smallpox DNA vaccine consisting of the A33R, A36R, L1R and B5R genes has been demonstrated to protect mice against vaccinia challenge (Hooper et al. 2003), and a DNA vaccine in combination with vaccination with these proteins protected Rhesus macaques against a lethal monkeypox virus challenge (Heraud et al. 2006). Although single antigens are effective in mice, the monkey model reveals that several antigens are required for protection against a poxvirus. It is likely that several antigens would be required for a protective immune response against capripoxviruses. Determining which antigens are the protective antigens will require capripoxvirus challenge experiments in the susceptible host species of sheep, goats and cattle. It is likely that the protective antigens would be the same for all capripoxviruses due to the similarities between these viruses.

The capripoxvirus antigens involved in T-cell immunity have not been characterized in sheep, goats or cattle. With vaccinia virus both CD4+ and CD8+ T-cell epitopes have been mapped in both mice and humans (Sette et al. 2009; Gilchuk et al. 2013). Identification of major histocompatibility class I antigens can be done by infecting cells with virus and then affinity purifying class I molecules eluting the peptides bound to the MHC I molecules and then identifying these peptides using mass spectrometry. Once these peptides have been identified, they can be validated by confirming that T-cell responses can be measured to these antigens. Bioinformatics can also be used to identify T-cell antigens for vaccinia (Moise et al. 2009). These approaches can be used to identify the numerous antigens involved for capripoxvirus infections.

References

Amara RR, Nigam P, Sharma S, Liu J, Bostik V (2004) Long-lived poxvirus immunity, robust CD4 help, and better persistence of CD4 than CD8 T cells. J Virol 78:3811–3816

Bowden TR, Coupar BE, Babiuk SL, White JR, Boyd V, Duch CJ, Shiell BJ, Ueda N, Parkyn GR, Copps JS, Boyle DB (2009) Detection of antibodies specific for sheeppox and goatpox viruses using recombinant capripoxvirus antigens in an indirect enzyme-linked immunosorbent assay. J Virol Methods 161:19–29

Bray M, Wright ME (2003) Progressive vaccinia. Clin Infect Dis 36:766–774

Chaudhri G, Panchanathan V, Bluethmann H, Karupiah G (2006) Obligatory requirement for antibody in recovery from a primary poxvirus infection. J Virol 80:6339–6344

Crotty S, Felgner P, Davies H, Glidewell J, Villarreal L, Ahmed R (2003) Cutting edge: long-term B cell memory in humans after smallpox vaccination. J Immunol 171:4969–4973

Edghill-Smith Y, Golding H, Manischewitz J, King LR, Scott D, Bray M, Nalca A, Hooper JW, Whitehouse CA, Schmitz JE, Reimann KA, Franchini G (2005) Smallpox vaccine-induced antibodies are necessary and sufficient for protection against monkeypox virus. Nat Med 11:740–747

Galmiche MC, Goenaga J, Wittek R, Rindisbacher L (1999) Neutralizing and protective antibodies directed against vaccinia virus envelope antigens. Virology 254:71–80

Gilchuk P, Spencer CT, Conant SB, Hill T, Gray JJ, Niu X, Zheng M, Erickson JJ, Boyd KL, McAfee KJ, Oseroff C, Hadrup SR, Bennink JR, Hildebrand W, Edwards KM, Crowe JE Jr, Williams JV, Buus S, Sette A, Schumacher TN, Link AJ, Joyce S (2013) Discovering naturally processed antigenic determinants that confer protective T cell immunity. J Clin Invest 123:1976–1987

Gordon J, Mohandas A, Wilton S, Dales S (1991) A prominent antigenic surface polypeptide involved in the biogenesis and function of the vaccinia virus envelope. Virology 181:671–686

Hammarlund E, Lewis MW, Hansen SG, Strelow LI, Nelson JA, Sexton GJ, Hanifin JM, Slifka MK (2003) Duration of antiviral immunity after smallpox vaccination. Nat Med 9:1131–1137

Hathaway WE, Githens JH, Blackburn WR, Fulginiti V, Kempe CH (1965) Aplastic anemia, histiocytosis and erythrodermia in immunologically deficient children: probable human runt disease. N Engl J Med 273:953–958

Heraud JM, Edghill-Smith Y, Ayala V, Kalisz I, Parrino J, Kalyanaraman VS, Manischewitz J, King LR, Hryniewicz A, Trindade CJ, Hassett M, Tsai WP, Venzon D, Nalca A, Vaccari M, Silvera P, Bray M, Graham BS, Golding H, Hooper JW, Franchini G (2006) Subunit recombinant vaccine protects against monkeypox. J Immunol 177:2552–2564

Hooper JW, Custer DM, Thompson E (2003) Four-gene-combination DNA vaccine protects mice against a lethal vaccinia virus challenge and elicits appropriate antibody responses in nonhuman primates. Virology 306:181–195

Hsiao JC, Chung CS, Chang W (1999) Vaccinia virus envelope D8L protein binds to cell surface chondroitin sulfate and mediates the adsorption of intracellular mature virions to cells. J Virol 73:8750–8761

Kitching RP (1986) Passive protection of sheep against capripoxvirus. Res Vet Sci 41:247–250

Moise L, McMurry JA, Buus S, Frey S, Martin WD, De Groot AS (2009) In silico-accelerated identification of conserved and immunogenic variola/vaccinia T-cell epitopes. Vaccine 27:6471–6479

Moutaftsi M, Tscharke DC, Vaughan K, Koelle DM, Stern L, Calvo-Calle M, Ennis F, Terajima M, Sutter G, Crotty S, Drexler I, Franchini G, Yewdell JW, Head SR, Blum J, Peters B, Sette A (2010) Uncovering the interplay between CD8, CD4 and antibody responses to complex pathogens. Future Microbiol 5:221–239

Ngichabe CK, Wamwayi HM, Ndungu EK, Mirangi PK, Bostock CJ, Black DN, Barrett T (2002) Long term immunity in African cattle vaccinated with a recombinant capripox-rinderpest virus vaccine. Epidemiol Infect 128:343–349

Panchanathan V, Chaudhri G, Karupiah G (2006) Protective immunity against secondary poxvirus infection is dependent on antibody but not on CD4 or CD8 T-cell function. J Virol 80:6333–6338

Parker AK, Parker S, Yokoyama WM, Corbett JA, Buller RM (2007) Induction of natural killer cell responses by ectromelia virus controls infection. J Virol 81:4070–4079

Perrin LH, Zinkernagel RM, Oldstone MB (1977) Immune response in humans after vaccination with vaccinia virus: generation of a virus-specific cytotoxic activity by human peripheral lymphocytes. J Exp Med 146:949–969

Rodriguez JF, Janeczko R, Esteban M (1985) Isolation and characterization of neutralizing monoclonal antibodies to vaccinia virus. J Virol 56:482–488

Sette A, Grey H, Oseroff C, Peters B, Moutaftsi M, Crotty S, Assarsson E, Greenbaum J, Kim Y, Kolla R, Tscharke D, Koelle D, Johnson RP, Blum J, Head S, Sidney J (2009) Definition of epitopes and antigens recognized by vaccinia specific immune responses: their conservation in variola virus sequences, and use as a model system to study complex pathogens. Vaccine 27 (Suppl 6):G21–G26

Weiss KE (1968) Lumpy skin disease virus. Virol Monogr 3:111–131

Wolffe EJ, Vijaya S, Moss B (1995) A myristylated membrane protein encoded by the vaccinia virus L1R open reading frame is the target of potent neutralizing monoclonal antibodies. Virology 211:53–63

Epidemiology and Transmission

13

Eyal Klement

13.1 Transmission Modes of LSDV

Pathogens can be either transmitted directly, through contact between animals, or indirectly, by fomites or vectors. The direct transmission of lumpy skin disease virus (LSDV) between animals is rare, suggesting that it is primarily transmitted by blood-sucking arthropod vectors. This conjecture was based on several field observations, which included the high abundance of biting arthropods during epidemics of LSD (Davies 1991; Yeruham et al. 1995), lack of transmission of the disease in insect proof cattle pens and sharp reduction in occurrence of LSD during periods of cold weather with frosts, attributed to a reduction in insect vector population (Davies 1991). Support for the indirect transmission mode of LSD was shown by a controlled study performed on Friesian-cross cattle (Carn and Kitching 1995). In this study seven experiments were performed in which two infected animals were housed together with a susceptible animal for 28 days and were followed for the development of LSD typical clinical signs. In each of these experiments, at least one of the infected animals suffered either from generalized LSD clinical signs or from a severe localized lesion. However, no clinical signs were observed in the contact animals. Six of the seven in-contact animals showed no delayed-type hypersensitivity reaction to intradermal challenge at 28 days and were fully susceptible to subsequent challenge. Additional support was shown in a study performed during an LSD outbreak in a dairy farm in the south of Israel in 2006. The location of each animal in ten pens in the farm was recorded daily, and a mathematical model based on these data showed that indirect transmission is sufficient to explain the transmission of the virus between the cattle in the herd, while direct transmission had a negligible role, if any, in the spread of the virus (Magori-Cohen et al. 2012).

13.2   Potential Vectors of Lumpy Skin Disease Virus

The indirect transmission of LSDV is assumed to be mechanically vectored by arthropods, in which the virus does not replicate or circulate. The virus' high stability makes it possible to survive in many different vectors. Mechanical virus transmission rate is inversely proportional to the virus survival in the interval between the vector blood meals. True flies (Diptera) with an interrupted or repeated feeding pattern can thus be efficient vectors of viruses. This feeding pattern, where blood meal taken after one bite is not sufficient, due to interruption by the host reaction, is forcing the vector to visit the same or different host in a short time, which is sufficient for the virus survival (Carn 1996). The role of an arthropod as a vector of LSDV should be demonstrated both on its competence and capacity. Vector competence is usually studied under controlled conditions and defines its ability to infect a susceptible animal after feeding on an infectious animal. Vector capacity summarizes quantitatively the basic biological and ecological attributes of the vector which are associates with viral transmission. These include traits like biting rate, feeding preference and frequency and the size of vector population (Reisen 2009).

To date there is no particular arthropod for which both competence and capacity were demonstrated. Vector competence of several arthropods was tested in the laboratory. *Aedes aegypti* female mosquitoes that had fed upon lesions of LSDV-infected cattle were able to transmit virus to susceptible cattle over a period of 2–6 days post-infective feeding. The virus was isolated from all the recipient steers, though only five of them developed disease which was usually mild (Chihota et al. 2001). *Aedes aegypti* is therefore a competent vector of LSDV. However, outbreaks of LSDV occurred in several European and Middle East countries, in which this vector is not abundant (Kraemer et al. 2015a, b). In the same genus, *Aedes albopictus* is also known as a competent vector of many viruses and although it is more widespread than *Ae. aegypti*, its presence in several affected countries was anecdotal during the eruption of LSD epidemics (Kraemer et al. 2015a). These mosquitoes prefer human blood over blood of other mammals (Lounibos and Kramer 2016), further reducing their capacity as vectors of LSDV.

The competence of the mosquitoes *Anopheles stephensi* and *Culex quinquefasciatus*, the stable fly *Stomoxys calcitrans* and the biting midge *Culicoides nubeculosus* were assessed as well. None of these blood-feeding dipterans were able to infect susceptible cattle 24 h after feeding on blood infected by LSDV. LSDV was identified from all of the dipteran species tested in excess of the minimum infectious dose for cattle via the intradermal and intravenous routes. However, while the mosquitoes were culture test positive for LSDV up to 4 days after feeding, *S. calcitrans* and *C. nubeculosus* were positive only on the feeding day (Chihota et al. 2003). Despite the failure of S. calcitrans to transmit LSDV in the above study, there are several evidences to support its vectoring potential. *Stomoxys calcitrans* was shown to transmit several animal pathogens including viruses (Baldacchino et al. 2013), and most importantly it was shown to transmit the Yemen capripox strain to a susceptible goat (Mellor et al. 1987). Both male and female flies feed on blood. It is an interrupted feeder which is mostly abundant near the legs of cattle and horses and can take 2–3 blood meals every day (Carn 1996). In a study performed to reveal the seasonal pattern of potential dipteran vectors of LSDV in dairy farms, the relative abundance of *S. calcitrans* in affected dairy farms was highest in December, January and April and was highly correlated with the occurrence of LSDV outbreaks. The abundance of other blood-feeding dipterans (e.g. biting midges and mosquitoes), however, was poorly associated with the timing of the outbreaks (Kahana-Sutin et al. 2016). Grazing beef cattle, during these outbreaks, was mostly affected during the summer months. It was therefore suggested that different flies might serve as vectors in grazing and zero-grazing herds. As the abundance of the horn fly *Haematobia irritans* was reported to be high in beef herds during the outbreaks, it was suggested as the potential vector in these settings (Kahana-Sutin et al. 2016). The circumstantial nature of this evidence, as well as the lack of successful transmission of other viruses by this fly, suggest that further studies are necessary before incriminating the horn fly as a potential vector of LSDV (Buxton et al. 1985; Chamorro et al. 2011).

Several studies have also demonstrated the competence of ticks as vectors of LSDV. Transstadial and mechanical transmission of LSDV was demonstrated in males of *Rhipicephalus appendiculatus* and *Amblyomma hebraeum*. This was demonstrated by both virus isolation from the saliva of fed adults, or from adults fed as nymphs on infected cattle, and by transmission of the virus by these ticks to susceptible animals (Lubinga et al. 2013, 2015; Tuppurainen et al. 2013b). Transovarial transmission of LSDV was shown in *R. decoloratus* ticks. The adults which developed from eggs laid by infected ticks infected susceptible cattle and caused viremia and mild clinical disease (Tuppurainen et al. 2013a). These evidences suggest that ticks may play an important role as reservoirs of LSDV. Their role in transmission of the virus in large outbreaks of cattle awaits capacity studies and is probably less important as large outbreaks have occurred in zero-grazing cattle where ticks are mostly rare (Yeruham et al. 1995; Magori-Cohen et al. 2012; Ben-Gera et al. 2015) and as the spread velocity of LSDV epidemics cannot be explained by tick borne transmission.

13.3 Direct Transmission of Lumpy Skin Disease

Observation of low levels of LSDV shedding in oral and nasal secretions between 12 and 18 days postinfection (Babiuk et al. 2008) may support the report of LSDV transmission to naïve cattle that was allowed to share drinking troughs with severely infected animals in insect-free facilities (Haig 1957). Transmission of LSDV to suckling calves through milk was suggested (Weiss 1968) but was never demonstrated in insect proof settings.

LSDV was found to be excreted in semen of bulls, starting from 8 days postinfection and up to 159 days. Virus isolation, however, was successful only in the case of occurrence of severe disease. The exact amount of virus could not be determined due to cytotoxicity, though virus isolation was successful from semen diluted 1:1000. Other bulls which were infected during this study and developed only mild clinical disease or only fever were positive by at least one PCR sample collected from semen, but with no successful virus isolation (Irons et al. 2005). In another study, insemination of naïve heifers with bull semen spiked with 5.5 log TCID50/ml of LSDV resulted in infection that succumbed in some of the animals to severe disease. Virus was also isolated from embryos of infected heifers. All of the infected heifers aborted very early after insemination, and up to 28 days post-insemination, none were pregnant (Annandale et al. 2014). This study shows that such LSDV transmission by semen is possible; nevertheless, the concentration of virus used in this study was high when compared to the low concentration of virus found in secretions of infected animals (Babiuk et al. 2008). Possible virus transmission by semen can be prevented using a good vaccine, as bulls vaccinated with the attenuated Neethling vaccine did not excrete the virus in semen despite the occurrence of mild clinical disease (Osuagwuh et al. 2007).

13.4   Transmission Via Subclinically Infected Cattle

Several studies demonstrated the occurrence of subclinical infection in cattle (Carn and Kitching 1995; Tuppurainen et al. 2005, 2013b). Transmission of LSDV by ticks that were fed on normal healthy looking skin (Tuppurainen et al. 2013b) suggest a potential role of subclinically infected animals in LSDV transmission. Transmission via subclinically infected animals could potentially occur either by ticks or by contact with intact skin. However, the level of viable virus in intact skin was undetectable in cattle with mild to moderate clinical disease (Babiuk et al. 2008). Virus isolation from blood is intermittent, short term and around five orders of magnitude lower titers compared to skin lesions (Tuppurainen et al. 2005; Babiuk et al. 2008). Additionally, outbreaks in Israel were controlled using modified stamping out (i.e. only stamping-out cases with generalized skin lesions), which were performed under the assumption that the high levels of virus present in the skin lesions is the most likely source of virus transmission (AHAW 2015), concurrently with the use of the original RM-65 sheeppox attenuated vaccine which was later shown to have very low effectiveness for preventing LSD (Ben-Gera et al. 2015). Taken together, the evidences gathered to date suggest that although subclinically infected animals may transmit the virus through bites of mechanical arthropod vectors, this route of transmission plays only a minor role in transmission of LSDV.

13.5   Spread of Lumpy Skin Disease

The median spread rate of LSD in the Balkans during the outbreaks of 2015–2016 was 7.3 km/week. However, the distribution was highly skewed to the right, with a maximum value reaching 543.6 km/week (Mercier et al. 2017). This suggests that similarly to the data described for the spread of bluetongue virus (Hendrickx et al. 2008), there are several modes of spread of lumpy skin disease. The most abundant mode is of short distance (up to few kilometres), which may be mostly driven by active flight of the insect vectors. This is accompanied by a significantly lower frequency of intermediate and long-term spread of up to few hundreds of kilometres. Cases of long-distance spread are most probably related to animal movements. This is supported by the results of epidemiological studies performed in Ethiopia, which show that the introduction of a new animal into a herd is a significant risk factor for the occurrence of LSD (Gari et al. 2010; Hailu et al. 2014). Long-distance spread of LSDV might be also attributed to wind borne movement of vectors as was previously suggested for LSDV (Klausner et al. 2015), as well as for the transmission of other vector-borne viruses of cattle-like bovine ephemeral fever (Aziz-Boaron et al. 2012), epizootic haemorrhagic disease virus (Kedmi et al. 2010) and bluetongue virus (Hendrickx et al. 2008).

13.6 Seasonality

The occurrence of LSD outbreaks is characterized by marked seasonality both in tropical and temperate regions. Usually it is associated with periods of high rainfall accompanied by warm temperatures and a resultant high insect activity (Hunter and Wallace 2001). Studies in Ethiopia reported three peaks of the disease, the highest occurred during August and two additional peaks were observed in May and December (Hailu et al. 2014), and an opposing results of highest incidence between September and December, and the lowest incidence in May (Ayelet et al. 2014) . In Turkey, seasonality was also observed, with two high peaks of the outbreaks during September and November and a smaller peak during March (Sevik and Dogan 2016). In Israel the highest peak was observed during April, followed by smaller peaks in August and December (Kahana-Sutin et al. 2016). In this study the authors also differentiated between peaks of the disease during the dry summer months, which occurred mostly in grazing beef herds and the peaks in the beginning of the winter and in the spring which occurred mostly in dairy herds, parallel to a rise of the stable fly population. In all of the above-mentioned countries, LSD occurred all over the year despite the seasonal changes in disease incidence. In the Balkans, probably due to the lower temperatures during the winter, the disease was absent during January to March and peaked during the summer months (Mercier et al. 2017).

13.7 Geographical Risk Factors

LSD was documented in a wide geographical range, in Africa, the Middle East, Central Asia and Europe and in a diversity of climates. Using the presence-only maximum entropy ecological niche modelling technique (Maxent) in order to characterize the geographical risk factors for disease occurring in the Middle East (Alkhamis and VanderWaal 2016), it was shown that annual precipitation (positive association) or/and mean diurnal temperature range (negative association) were the most significant environmental factors to be associated with LSD outbreak distribution. This supports the notion that humid and warm regions are most appropriate for the development of LSD outbreaks as they support the vector population. Despite this, LSD outbreaks can occur in moderate temperatures of 18–22 °C, as occurred in the Evros region in Greece (Tasioudi et al. 2016).

In the Maxent model based on the Israeli data, land cover of croplands and urban and mixed rain-fed arid livestock production systems were also found as a significant risk factors for the occurrence of LSD outbreaks (Alkhamis and VanderWaal 2016). In Turkey during 2014–2015, proximity to a lake was associated with a 1.5 higher risk for occurrence of an LSD case (Sevik and Dogan 2016).

In Africa LSD has occurred in all the diverse ecological zones, from the high altitude temperate grasslands, the wet and dry bushed and the wooded savannah to the dry semi-desert (Davies 1991). In Ethiopia and Zimbabwe, the highest incidence of LSD was documented in regions characterized by moist, humid conditions and the presence of flooding and irrigation (Hailu et al. 2014; Gomo et al. 2017), though the disease is abundant in almost all regions and agroecological zones of the country (Ayelet et al. 2014).

These studies support the hypothesis of various mechanical arthropod vectors for LSDV.

13.8   Risk Factors in the Herd Level

The protection conferred by herd vaccination is discussed in detail in other parts of this book. The influence of other factors at the herd level was examined in several studies, which were mainly performed in Africa. The caveat of most of these studies is the poor control for various confounders, such as region, climatic factors and vaccination. Hence the results are quiet inconsistent. In a study performed in Ethiopia, herd size was found to be positively associated with the risk for LSD (Hailu et al. 2014). The same association was found in Turkey (Sevik and Dogan 2016). It should be remembered, however, that larger herds have higher probability for having at least one case of LSD, on a purely chance. Thus, it is not necessary that large herds are more exposed to the virus or more susceptible to infection. Other risk factors found in this study are communal watering and grazing, agro-pastoral farming systems and new cattle introduction. In another study performed in Ethiopia, feedlot cattle was found to be in higher risk for LSD infection compared to extensively managed herds (Ayelet et al. 2014). In Turkey, the incidence in beef herds was higher than in dairy herds, though this difference was not statistically significant (Sevik and Dogan 2016). In Zimbabwe, LSD morbidity was highest in resettlement farms, though the authors explain this finding by higher accessibility of veterinary service in these regions (Gomo et al. 2017).

13.9   Risk Factors in the Animal Level

Breeds of zebu type indigenous to Africa are generally less susceptible to infection by LSD and may develop extensive skin lesions but have less severe clinical disease and lower mortality rates than cattle exotic to Africa (Davies 1991). Similar findings were reported in studies conducted in Ethiopia, Turkey and the Sultanate of Oman (Gari et al. 2010; Tageldin et al. 2014; Sevik and Dogan 2016). In these studies a more severe disease and a higher mortality were observed in European cross breeds, as compared to local breeds. Interestingly, in a study conducted in Ethiopia, similar morbidity rates were observed in zebu cattle and zebu-Holstein cross breeds. However, in the zebu cattle morbidity rate among vaccinated, cattle was more than four times higher than among non-vaccinated, while in the cross breeds, vaccine did not show any protective effect. These findings might be the result of nonstandardized definition of morbidity and lack of control for various confounding effects (Ayelet et al. 2013).

There are contradicting results regarding age-related susceptibility to LSD. While in some studies, higher morbidity was observed in young animals (Ayelet et al. 2013, 2014), others showed no association of morbidity with age (Sevik and Dogan 2016) or even lower morbidity among calves (Magori-Cohen et al. 2012). The same inconsistency is reported also for the effect of sex on susceptibility to LSD (Magori-Cohen et al. 2012; Ayelet et al. 2013, 2014).

13.10 Further Suggested Studies

There is still a significant lack of data regarding several epidemiological characteristics of LSD. While it is highly evident that the virus is mechanically transmitted by arthropod vectors, the most important vectors in the different geographical regions should still be defined. It is thus very important that data from vector competence and vector capacity studies will be integrated. With this, it is still prudent to assess the exact impact of direct virus transmission from animal to animal as well as to quantify the role of subclinically infected animals in the spread of the virus. Natural transmission by semen was not demonstrated yet though studies show that this may be possible. Risk factors for LSD are based on observational studies, which might have been influenced by significant confounding. More epidemiological studies are needed and performed with better methodology. Increasing the knowledge regarding these aspects will provide better data for the performance of accurate risk assessment regarding LSD and will help in defining the best strategies for disease control. Experiments are required to determine the minimal dose of lumpy skin disease virus required to start an infection by different vectors and ticks. Complicating matters are the enveloped and mature virus particle forms which may have different levels of infectivity as well as the site of initial replication with the skin, specifically the keratinocytes being the most susceptible cells. In addition, the role of insect vectors or tick saliva as well as the trauma caused by feeding may also play a role in promoting infection in the skin.

References

Alkhamis MA, VanderWaal K (2016) Spatial and temporal epidemiology of lumpy skin disease in the Middle East, 2012–2015. Front Vet Sci 3:19

Annandale CH, Holm DE, Ebersohn K, Venter EH (2014) Seminal transmission of lumpy skin disease virus in heifers. Transbound Emerg Dis 61:443–448

Ayelet G, Abate Y, Sisay T, Nigussie H, Gelaye E, Jemberie S, Asmare K (2013) Lumpy skin disease: preliminary vaccine efficacy assessment and overview on outbreak impact in dairy cattle at Debre Zeit, central Ethiopia. Antiviral Res 98:261–265

Ayelet G, Haftu R, Jemberie S, Belay A, Gelaye E, Sibhat B, Skjerve E, Asmare K (2014) Lumpy skin disease in cattle in central Ethiopia: outbreak investigation and isolation and molecular detection of the virus. Revue Sci Tech 33:877–887

Aziz-Boaron O, Klausner Z, Hasoksuz M, Shenkar J, Gafni O, Gelman B, David D, Klement E (2012) Circulation of bovine ephemeral fever in the Middle East—strong evidence for transmission by winds and animal transport. Vet Microbiol 158:300–307

Babiuk S, Bowden TR, Parkyn G, Dalman B, Manning L, Neufeld J, Embury-Hyatt C, Copps J, Boyle DB (2008) Quantification of lumpy skin disease virus following experimental infection in cattle. Transbound Emerg Dis 55:299–307

Baldacchino F, Muenworn V, Desquesnes M, Desoli F, Charoenviriyaphap T, Duvallet G (2013) Transmission of pathogens by Stomoxys flies (Diptera, Muscidae): a review. Parasite 20:26

Ben-Gera J, Klement E, Khinich E, Stram Y, Shpigel NY (2015) Comparison of the efficacy of Neethling lumpy skin disease virus and x10RM65 sheep-pox live attenuated vaccines for the prevention of lumpy skin disease: the results of a randomized controlled field study. Vaccine 33:4837–4842

Buxton BA, Hinkle NC, Schultz RD (1985) Role of insects in the transmission of bovine leukosis virus: potential for transmission by stable flies, horn flies, and tabanids. Am J Vet Res 46:123–126

Carn VM (1996) The role of dipterous insects in the mechanical transmission of animal viruses. Br Vet J 152:377–393

Carn VM, Kitching RP (1995) An investigation of possible routes of transmission of lumpy skin disease virus (Neethling). Epidemiol Infect 114:219–226

Chamorro MF, Passler T, Givens MD, Edmondson MA, Wolfe DF, Walz PH (2011) Evaluation of transmission of bovine viral diarrhea virus (BVDV) between persistently infected and naive cattle by the horn fly (Haematobia irritans). Vet Res Commun 35:123–129

Chihota CM, Rennie LF, Kitching RP, Mellor PS (2001) Mechanical transmission of lumpy skin disease virus by Aedes aegypti (Diptera: Culicidae). Epidemiol Infect 126:317–321

Chihota CM, Rennie LF, Kitching RP, Mellor PS (2003) Attempted mechanical transmission of lumpy skin disease virus by biting insects. Med Vet Entomol 17:294–300

Davies FG (1991) Lumpy skin disease, an African capripox virus disease of cattle. Br Vet J 147:489–503

EFSA AHAW Panel (EFSA Panel on Animal Health and Welfare) (2015) Scientific opinion on lumpy skin disease. EFSA J 13(1):3986, 73 pp. https://doi.org/10.2903/j.efsa.2015.3986

Gari G, Waret-Szkuta A, Grosbois V, Jacquiet P, Roger F (2010) Risk factors associated with observed clinical lumpy skin disease in Ethiopia. Epidemiol Infect 138:1657–1666

Gomo C, Kanonhuwa K, Godobo F, Tada O, Makuza SM (2017) Temporal and spatial distribution of lumpy skin disease (LSD) outbreaks in Mashonaland West Province of Zimbabwe from 2000 to 2013. Trop Anim Health Prod 49:509–514

Haig DA (1957) Lumpy skin disease. Bull Epizoot Dis Afr:421–430

Hailu B, Tolosa T, Gari G, Teklue T, Beyene B (2014) Estimated prevalence and risk factors associated with clinical lumpy skin disease in north-eastern Ethiopia. Prev Vet Med 115:64–68

Hendrickx G, Gilbert M, Staubach C, Elbers A, Mintiens K, Gerbier G, Ducheyne E (2008) A wind density model to quantify the airborne spread of Culicoides species during north-western Europe bluetongue epidemic, 2006. Prev Vet Med 87:162–181

Hunter P, Wallace D (2001) Lumpy skin disease in southern Africa: a review of the disease and aspects of control. J S Afr Vet Assoc 72:68–71

Irons PC, Tuppurainen ES, Venter EH (2005) Excretion of lumpy skin disease virus in bull semen. Theriogenology 63:1290–1297

Kahana-Sutin E, Klement E, Lensky I, Gottlieb Y (2016) High relative abundance of the stable fly Stomoxys calcitrans is associated with lumpy skin disease outbreaks in Israeli dairy farms. Med Vet Entomol 31:150–160

Kedmi M, Herziger Y, Galon N, Cohen RM, Perel M, Batten C, Braverman Y, Gottlieb Y, Shpigel N, Klement E (2010) The association of winds with the spread of EHDV in dairy cattle in Israel during an outbreak in 2006. Prev Vet Med 96:152–160

Klausner Z, Fattal E, Klement E (2015) Using synoptic systems' typical wind trajectories for the analysis of potential atmospheric long distance dispersal of lumpy skin disease virus. Transbound Emerg Dis 64:398–410

Kraemer MU, Sinka ME, Duda KA, Mylne A, Shearer FM, Brady OJ, Messina JP, Barker CM, Moore CG, Carvalho RG, Coelho GE, Van Bortel W, Hendrickx G, Schaffner F, Wint GR, Elyazar IR, Teng HJ, Hay SI (2015a) The global compendium of Aedes aegypti and Ae. albopictus occurrence. Sci Data 2:150035

Kraemer MU, Sinka ME, Duda KA, Mylne AQ, Shearer FM, Barker CM, Moore CG, Carvalho RG, Coelho GE, Van Bortel W, Hendrickx G, Schaffner F, Elyazar IR, Teng HJ, Brady OJ, Messina JP, Pigott DM, Scott TW, Smith DL, Wint GR, Golding N, Hay SI (2015b) The global distribution of the arbovirus vectors Aedes aegypti and Ae. albopictus. Elife 4:e08347

Lounibos LP, Kramer LD (2016) Invasiveness of Aedes aegypti and Aedes albopictus and vectorial capacity for Chikungunya Virus. J Infect Dis 214:S453–S458

Lubinga JC, Tuppurainen ES, Stoltsz WH, Ebersohn K, Coetzer JA, Venter EH (2013) Detection of lumpy skin disease virus in saliva of ticks fed on lumpy skin disease virus-infected cattle. Exp Appl Acarol 61:129–138

Lubinga JC, Tuppurainen ES, Mahlare R, Coetzer JA, Stoltsz WH, Venter EH (2015) Evidence of transstadial and mechanical transmission of lumpy skin disease virus by Amblyomma hebraeum ticks. Transbound Emerg Dis 62:174–182

Magori-Cohen R, Louzoun Y, Herziger Y, Oron E, Arazi A, Tuppurainen E, Shpigel NY, Klement E (2012) Mathematical modelling and evaluation of the different routes of transmission of lumpy skin disease virus. Vet Res 43(1):1

Mellor PS, Kitching RP, Wilkinson PJ (1987) Mechanical transmission of capripox virus and African swine fever virus by Stomoxys calcitrans. Res Vet Sci 43:109–112

Mercier A, Arsevska E, Bournez L, Bronner A, Calavas D, Cauchard J, Falala S, Caufour P, Tisseuil C, Lefrancois T, Lancelot R (2017) Spread rate of lumpy skin disease in the Balkans, 2015–2016. Transbound Emerg Dis 65:240–243

Osuagwuh UI, Bagla V, Venter EH, Annandale CH, Irons PC (2007) Absence of lumpy skin disease virus in semen of vaccinated bulls following vaccination and subsequent experimental infection. Vaccine 25:2238–2243

Reisen WK (2009) Epidemiology of vector borne diseases. In: Mullen G, Durden L (eds) Medical and veterinary entomology. Elsevier, London

Sevik M, Dogan M (2016) Epidemiological and molecular studies on lumpy skin disease outbreaks in Turkey during 2014–2015. Transbound Emerg Dis 64:1268–1279

Tageldin MH, Wallace DB, Gerdes GH, Putterill JF, Greyling RR, Phosiwa MN, Al Busaidy RM, Al Ismaaily SI (2014) Lumpy skin disease of cattle: an emerging problem in the Sultanate of Oman. Trop Anim Health Prod 46:241–246

Tasioudi KE, Antoniou SE, Iliadou P, Sachpatzidis A, Plevraki E, Agianniotaki EI, Fouki C, Mangana-Vougiouka O, Chondrokouki E, Dile C (2016) Emergence of lumpy skin disease in Greece, 2015. Transbound Emerg Dis 63:260–265

Tuppurainen ES, Venter EH, Coetzer JA (2005) The detection of lumpy skin disease virus in samples of experimentally infected cattle using different diagnostic techniques. Onderstepoort J Vet Res 72:153–164

Tuppurainen ES, Lubinga JC, Stoltsz WH, Troskie M, Carpenter ST, Coetzer JA, Venter EH, Oura CA (2013a) Evidence of vertical transmission of lumpy skin disease virus in *Rhipicephalus decoloratus* ticks. Ticks and tick-borne diseases 4:329–333

Tuppurainen ES, Lubinga JC, Stoltsz WH, Troskie M, Carpenter ST, Coetzer JA, Venter EH, Oura CA (2013b) Mechanical transmission of lumpy skin disease virus by *Rhipicephalus appendiculatus* male ticks. Epidemiol Infect 141:425–430

Weiss KE (1968) Lumpy skin disease. Virol Monogr 3:111–131

Yeruham I, Nir O, Braverman Y, Davidson M, Grinstein H, Haymovitch M, Zamir O (1995) Spread of lumpy skin disease in Israeli dairy herds. Vet Rec 137:91–93

Part II

Early Detection of Lumpy Skin Disease, Diagnostic Tools and Treatment

Clinical Signs

14

Shawn Babiuk

The clinical signs caused by the lumpy skin disease virus (LSDV) are highly variable in the severity of disease in both natural outbreaks and in experimental inoculation of cattle. Clinical disease can range from unapparent, mild or moderate to severe form. The factors affecting the wide range of disease are likely to be complex and multifactorial, including the dose of virus inoculate, genetic factors of the host and the virus as well as the immune competence and possibly the age of the host with some studies demonstrating younger animals being more susceptible. With sheeppox and goatpox, younger animals are more susceptible. Clinical signs caused by LSDV were demonstrated to be much more severe in high-producing dairy breeds such as Holstein Friesian cattle compared to indigenous breeds (Davies 1991; Tageldin et al. 2014).

The disease is characterized first by fever ranging from 40°C to 41.5°C with lachrymation, inappetence, depression and unwillingness to move. Fever occurs around 5 days following experimental inoculation and remains elevated over several days. In the next few days of the onset of fever, eruption of skin lesions, so-called nodules occurs. These nodules range in size from 5 to 50 mm and are circular, raised, firm and well-circumscribed. Large irregular circumscribed plaques can occur from fused nodules (Fig. 14.1). The deep nodules are present throughout all layers of the skin, including the epidermis, dermis and adjacent subcutaneous layers and sometimes even the adjacent musculature. The clinical presentation of the skin lesions can vary dramatically in cattle with respect to numbers and size (Figs. 14.1, 14.2). These nodules may be painful and usually appear first around the head, including the mouth, nose (Fig. 14.3) and eyes, followed by the neck, body, udder, genitals (Fig. 14.4), legs and tail. The number of nodules in an infected animal can range from a single nodule to over a thousand in severely affected cattle. Later, the skin lesions often become necrotic plugs (Fig. 14.5) or so-called sitfast which then slough off, leaving large ulcers in the skin. These necrotic cores are very susceptible for secondary bacterial infections and are attractive for flies. When skin nodules heal, they leave permanent scars on the hides.

© Springer International Publishing AG, part of Springer Nature 2018
E. S. M. Tuppurainen et al., *Lumpy Skin Disease*,
https://doi.org/10.1007/978-3-319-92411-3_14

Fig. 14.1 Clinical picture of lumpy skin disease in cattle. Photo by Dr. Lior Zamir

Fig. 14.2 Cow with multiple skin nodules approximately 2 weeks postinfection. Photo by Dr. Eeva Tuppurainen

Cattle may become reluctant to move as a result of the enlarged lymph nodes and swelling of the brisket and legs. Lymphoid hyperplasia and generalized lymphade-nopathy as well as oedema frequently occur. Lameness may also occur due to skin lesions extending into the underlying tissue such as tendons and tendon sheaths. The

Fig. 14.3 Lumpy skin
disease infected bull showing
pox lesions in the muzzle.
Photo by Dr. Eeva
Tuppurainen

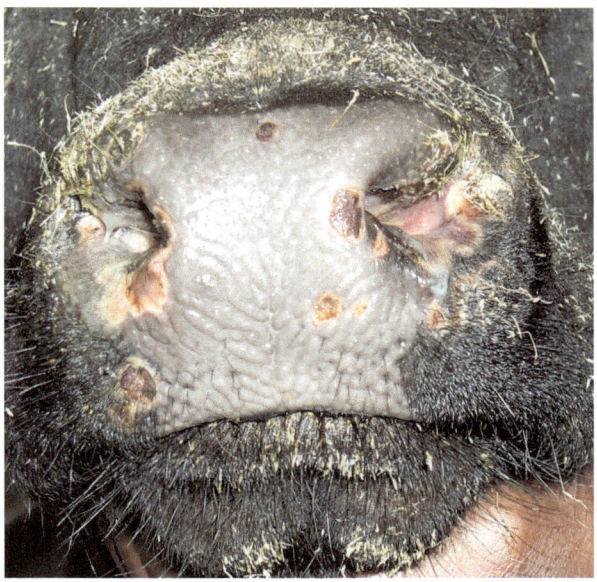

infected cattle can have increased levels in serum alanine aminotransferase, aspartate aminotransferase activities, creatinine level and creatinine phosphokinase in their blood (Neamat-Allah 2015; Şevik et al. 2016).

Rhinitis and nasal discharge starts as serous but later becomes mucopurulent. Conjunctivitis and ocular discharge can occur and sometimes keratitis is observed. In addition, excessive salivation, a loss of appetite leading to weight loss and depression may also occur. Characteristic pox lesions can develop in the mucous membranes of the mouth including the inside of the lips, gingivae and dental pads, tongue, soft palate, pharynx, epiglottis as well as the digestive tract. These lesions are likely the cause of inappetence and weight loss (Fig. 14.6). In addition, pox lesions can be found in the mucous membranes of the nasal cavities, turbinate, trachea and lungs. Infection in the lung can lead to primary or secondary pneumonia and respiratory distress.

Even though the case fatality of lumpy skin disease virus is low, the affected cattle become debilitated and can remain in poor condition for many months following infection. The scars destroy the value of the hide for use in the leather industry. Milk yield is reduced in lactating cattle and mastitis can occur. Abortions can occur in pregnant cattle, and there have been reports of aborted foetuses having multiple skin lesions as well as calves born with extensive skin lesions (Rouby and Aboulsoud 2016). A recent report describes a premature 1-day-old calf which was delivered by a cow that had lumpy skin disease in the seventh month of pregnancy. The calf died 36 h after birth and was weak, immature with a low body weight, ill-defined teeth, hyperaemic oral mucosa and respiratory distress. The calf had hard nodules on the skin and "sitfasts". Necropsy revealed nodules in the lungs, liver and ruminal pillars as well as enlarged lymph nodes (Rouby and Aboulsoud 2016).

Fig. 14.4 Lumpy skin
disease infected bull showing
skin lesions in the scrotum.
Photo by Dr. Eeva
Tuppurainen

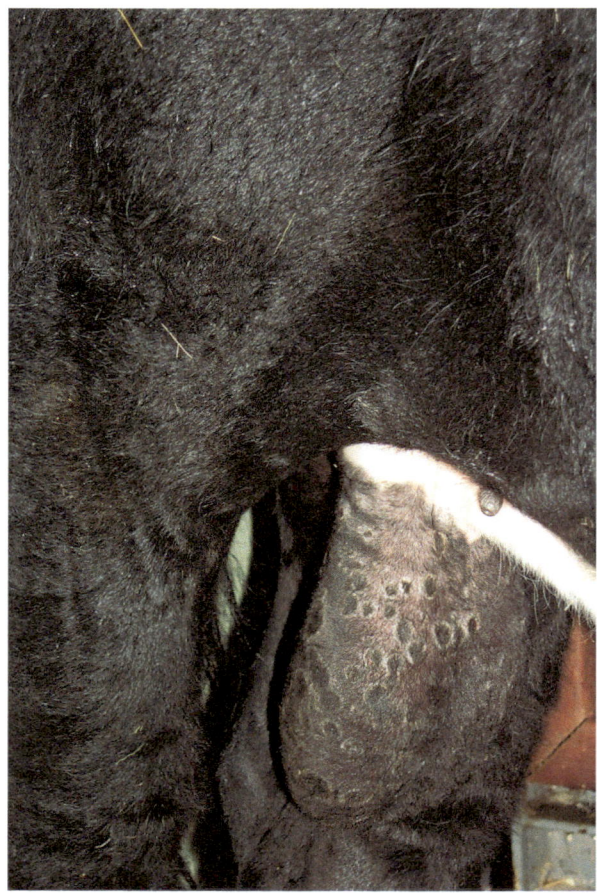

Fig. 14.5 Infected cow
showing lumpy skin lesions
with scabs at least 3 weeks
postinfection. Photo by
Dr. Lior Zamir

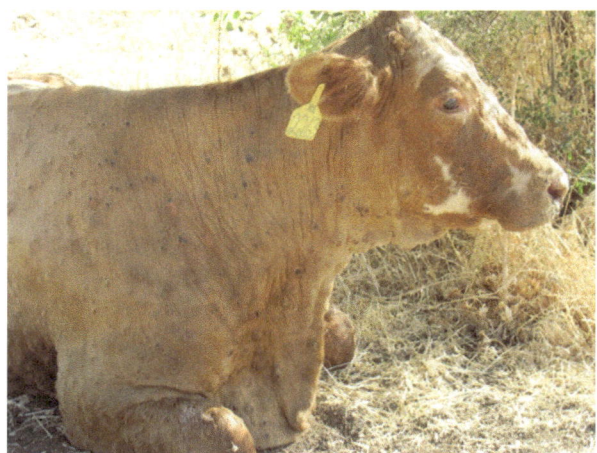

Fig. 14.6 Lumpy skin disease infected cow, showing a loss of body weight and skin lesions. Photo by Dr. Eeva Tuppurainen

During the LSD outbreak in 2006–2007 in Egypt, the ovarian activity in 640 cows was examined using ultrasonography. A high percentage of the LSD infected cows (93%) suffered from ovarian inactivity and showed no signs of oestrus, with smaller than average ovaries (Ahmed and Zaher 2008). In bulls, the scrotum, glans penis, preputial mucosa and parenchyma of the testes can be affected (Annendale et al. 2010).

References

Ahmed WM, Zaher KS (2008) Observations on lumpy skin disease in local Egyptian cows with emphasis on its impact on ovarian function. Afr J Microbiol Res 2:252–257

Annandale CH, Irons PC, Bagla VP, Osuagwuh UI, Venter EH (2010) Sites of persistence of lumpy skin disease virus in the genital tract of experimentally infected bulls. Reprod Domest Anim 45:250–255

Davies FG (1991) Lumpy skin disease, an African capripox virus disease of cattle. Br Vet J 147:489–503

Neamat-Allah AN (2015) Immunological, hematological, biochemical, and histopathological studies on cows naturally infected with lumpy skin disease. Vet World 8:1131–1136

Rouby S, Aboulsoud E (2016) Evidence of intrauterine transmission of lumpy skin disease virus. Vet J 209:193–195

Şevik M, Avci O, Doğan M, İnce ÖB (2016) Serum biochemistry of lumpy skin disease virus-infected cattle. Biomed Res Int 2016:6257984

Tageldin MH, Wallace DB, Gerdes GH, Putterill JF, Greyling RR, Phosiwa MN, Al Busaidy RM, Al Ismaaily SI (2014) Lumpy skin disease of cattle: an emerging problem in the Sultanate of Oman. Trop Anim Health Prod 46:241–246

Sample Collection and Transport

15

Shawn Babiuk

Sample collection should be carried out as described in the OIE Manual for Diagnostic Tests in the lumpy skin disease chapter (OIE 2017). The field manual for lumpy skin disease by the Food and Agriculture Organization of the United Nations (FAO) (Tuppurainen et al. 2017) provides practical guidance for the collection and safe transport of sample material to the local or international reference laboratories.

For live cattle, a biopsy of skin nodules contains high virus titres to be detected by, for example, highly sensitive PCR methods. However, these samples have to be collected aseptically using local anaesthesia. Dried scabs formed on top of the skin lesions are excellent sample material, and usually they are easy to pick up from infected animal. The virus is well protected inside the crust, and the scabs can be sent in a plain container/tube without any transport media into the laboratory. However, in case these scabs are found, the virus has probably been circulating within the herd already at least 3–4 weeks.

Nasal and saliva swabs should be collected in virus transport media (phosphate buffered saline with antibiotics) and stored at 4 °C, on ice or at −20 °C. The whole blood should be collected into ethylenediaminetetraacetic acid (EDTA)- or heparin-containing blood collection tubes stored at 4 °C, on ice or at −20 °C. Nasal and saliva swabs as well as whole blood can be used for virus isolation and molecular testing for lumpy skin disease. Sera should be collected in serum separation blood collection tubes and stored at 4 °C, on ice or at −20 °C for serological testing.

At post-mortem, skin lesions, lung lesions, lymph nodes as well as pox-like lesions on the internal organs should be collected. Both fresh and fixed samples should be collected with the fresh samples collected and stored at 4 °C, on ice or at −20 °C which are to be used for virus isolation and molecular testing for lumpy skin disease. Fixed samples collected in ten times the sample volume of 10% formalin and kept at ambient temperature.

© Springer International Publishing AG, part of Springer Nature 2018
E. S. M. Tuppurainen et al., *Lumpy Skin Disease*,
https://doi.org/10.1007/978-3-319-92411-3_15

References

OIE—World Organization for Animal Health (2017) Lumpy skin disease Chapter 2.4.13. OIE manual for diagnostic tests and vaccines for terrestrial animals

Tuppurainen E, Alexandrov T, Beltrán-Alcrudo D (2017) Lumpy skin disease field manual: a manual for veterinarians. FAO Animal Production and Health Manual No. 20. Food and Agriculture Organization of the United Nations (FAO), Rome, 60 p

Shawn Babiuk

Although the clinical disease presentation and the visceral pox lesions in cattle caused by lumpy skin disease virus (LSDV) are strongly indicative of lumpy skin disease (LSD), a definitive diagnosis requires laboratory confirmation. Milder forms of LSD can be confused with many different agents or diseases. Allergic reactions and physical trauma to the skin caused by insect and/or tick bites as well as urticaria and photosensitisation also need to be ruled out. These include differentials of several agents that cause skin lesions including viral agents such as parapoxviruses, bovine papular stomatitis virus and pseudocowpox, orthopoxviruses such as vaccinia and cowpox and bovine herpesvirus 2 causing pseudo lumpy skin disease. Since rinderpest has been eradicated (Roeder 2011), it is no longer a differential. Adverse reactions to LSDV vaccines can also occur (so-called Neethling disease) and is characterized by the appearance of skin nodules which are smaller than those caused by virulent LSDV field strain. Other skin diseases in cattle caused by bacterial agents such as hypoderma bovis infection, cutaneous tuberculosis, dermatophilosis and *Corynebacterium pseudotuberculosis* are also differentials. Additional differentials comprise demodicosis or mange caused by *Demodex bovis* as well as other skin lesions caused by parasites such as onchocercosis caused by *Onchocerca ochengi* or besnoitiosis caused by the protozoa *Besnoitia besnoiti*.

Since members of the *Capripoxvirus* genus have a general host tropism for their host species, before the molecular diagnostic methods become available, the virus was always classified according to the host it was isolated from. For example, if a capripoxvirus was isolated from a sheep, the virus would be called a sheeppox virus; if isolated from a goat, it would be goatpox; and if isolated from cattle, the virus would be classified as a LSDV. This has generally been useful although, it has caused some confusion with certain viruses. Generally LSDV does not cause disease in sheep and goats. However, there has been one instance where LSDV caused disease in sheep in Kenya. This virus was assumed to be a sheeppox virus; however, genetic sequencing of the Kenyan virus isolated has revealed that was LSDV (Tuppurainen et al. 2014). Therefore, it must be kept in mind that the general rule of classifying capripoxviruses based on the host species where the virus was isolated

E. S. M. Tuppurainen et al., *Lumpy Skin Disease*,
https://doi.org/10.1007/978-3-319-92411-3_16

is not perfect and additional molecular approaches should be used to confirm the virus identity.

Several different methods can be used for diagnosis. These include classical methods such as electron microscopy and virus isolation as well as more modern molecular methods including various PCR and real-time PCR, loop-mediated isothermal amplification (LAMP) and DNA sequencing protocols. Pathology and immunohistochemistry to observe capripoxvirus antigen in infected tissues can also be used. For serology, virus neutralization is the gold standard although there have been several ELISA's and immunostaining protocols developed.

Electron microscopy can be used to identify capripoxvirus in skin lesions. Due to the identical virus particle size between capripoxviruses, electron microscopy cannot be used to distinguish capripoxviruses into their specific virus (Kitching and Smale 1986). In addition, electron microscopy cannot differentiate capripoxviruses from orthopoxviruses without the application of specific immunological staining as the morphology of the brick-like structure and lateral bodies observed between capripoxviruses and othropoxviruses are identical. The cost of owning and maintaining electron microscopy capabilities and the development of more sensitive and rapid molecular protocols limit the practical use of electron microscopy in routine diagnostics.

Virus isolation can be used to isolate LSDV. Skin lesions contain very high levels of virus which can be easily isolated by first homogenizing the skin lesion and placing the clarified homogenate on susceptible cells. Cells are then be monitored daily for cytopathic effect. One downfall of virus isolation is the time that is required to isolate the virus as it can take up to 10 days to observe cytopathic effect and requires more than a single passage to isolate the virus. Capripoxvirus isolation can be confirmed by immunostaining using anti-capripoxvirus serum (Gulbahar et al. 2006; Babiuk et al. 2007). It is not possible to differentiate between sheeppox, goatpox and LSDV using cell culture since the cytopathic effect of these viruses is identical and there are no capripoxvirus serotypes (Kitching 1986) preventing specific antibody reagents to be developed. The length of time required to isolate LSDV is a limiting factor and the reason why molecular-based protocols are being used more for diagnostics.

Many different conventional, real-time and LAMP molecular assays have been developed for the detection of capripoxviruses. Three conventional PCR assays were developed using the P32 gene as a target (Ireland and Binepal 1998; Heine et al. 1999; Mangana-Vougiouka et al. 1999).

A multiplex PCR-based species-specific primer to differentiate between capripoxvirus species has been developed (Orlova et al. 2006). A duplex PRC assay was developed to detect both capripoxvirus and Orf virus using the A29L gene region of capripoxvirus (Zheng et al. 2007). Although this assay was only evaluated on sheeppox and goatpox viruses, the capripoxvirus primers in the assay will also amplify LSDV based on sequence homology.

Two different real-time PCRs have been developed to detect the three capripoxvirus members. The Balinsky assay is based on detecting ORF 068 [poly (A) polymerase (small subunit) gene] (Balinsky et al. 2008), whereas the Bowden assay is based the P32 region (Bowden et al. 2008). The Bowden real-time PCR assay has been further validated for use to OIE standards (Stubbs et al. 2012). Although these real-time PCR assays are very sensitive, unfortunately these assays are not able to differentiate between the three different capripoxvirus species.

Since members of the capripoxvirus genus are genetically highly similar, different genes have been assessed to determine if these genes can be used to categorize capripoxviruses as a sheeppox, goatpox or LSDV. The P32 gene has been demonstrated to be able to characterize sheeppox and goatpox viruses. The RPO30 gene has been identified as a gene that can be used to classify capripoxviruses into sheeppox, goatpox and LSD although there were some exceptions where a sheeppox was classified as a goatpox virus and where goatpox viruses were characterized as sheeppox viruses (Le Goff et al. 2009). A unique 21-nucleotide deletion present in the RPO30 gene in sheeppox but not in goatpox or LSD isolates allowed for a PCR assay to be developed to amplify a 151 bp fragment for sheeppox and a 172 bp fragment for goatpox as well as LSDV (Lamien et al. 2011a). These size differences of the PCR products can be observed by running them on an agarose gel to determine if the virus is a sheeppox or either a goatpox or LSDV (Lamien et al. 2011a). Building on this work, a real-time PCR assay was developed to be able to amplify all capripoxvirus members but to be able to differentiate them based on nucleotide differences observed in the G-protein-coupled chemokine receptor (GPCR) gene. A probe was designed so that the sequence differences between capripoxvirus members would lead to different melting temperatures for the probe due to mismatches. The probe had a Tm of 52 °C for SPPV due to five mismatches, a Tm of 61 °C for LSDV due to three mismatches and Tm of 69 °C for GTPV which matches 100% to the acceptor probe. These Tm differences were then observed using fluorescence melting curve analysis (FMCA) (Lamien et al. 2011b). To decrease the cost of this assay, unlabelled snapback primers in the presence of dsDNA intercalating EvaGreen dye were used to amplify a region of 96 bp within the CaPVs RPO30 gene and characterize the capripoxvirus using melting point analysis (Gelaye et al. 2013). A further improvement is the development of a pan-pox assay to simultaneously identify cowpox, camelpox, sheeppox, goatpox, LSD, orf, pseudocowpox and bovine papular stomatitis virus using high-resolution melting curve analysis of PCR amplicons (Gelaye et al. 2017).

A loop-mediated isothermal amplification assay (LAMP) has been developed targeting a conserved gene encoding the poly(A) polymerase small subunit (Das et al. 2012) An additional lamp assay was developed to the conserved region of the CaPV P32 gene (Murray et al. 2013). A lamp assay has been developed to distinguish sheeppox from goatpox by using two different sets of lamp primers (Zhao et al. 2014). This can likely be expanded to include a lamp primer set for specific detection of LSDV. Advantages of LAMP assays are they are user-friendly, simple to use, inexpensive and highly sensitive, making them suitable for the diagnosis of capripox in resource-constrained areas. A recombinase polymerase amplification assay has been developed for LSDV (Shalaby et al. 2016). The benefit of this assay is the rapid speed of only 15 min required for the assay.

If vaccination is being used to control the disease, additional PCR-based tests are required to differentiate between the vaccine and the wild-type LSDV due to mild but typical LSD symptoms appearing in a few vaccinated animals and diagnostic testing being positive for LSD (Menasherow et al. 2014). The differentiation between the LSD vaccine and wild-type LSDV is possible due to a known genetic difference of a 27 base pair (bp) fragment of the LSDV126 extracellular enveloped virus (EEV) gene that is present in field viruses but is absent from the Neethling LSD vaccine strains. A nested gradient PCR used together with restriction fragment polymorphism RFLP using MboI restriction enzyme which is specific for the vaccine isolate has been developed (Menasherow et al. 2014). Although this assay is useful, it was not user-friendly as it required several steps, leading to the development of a high-resolution melting (HRM) assay (Menasherow et al. 2016). This HRM assay is based on the temperature difference of ∼0.5 °C melting point change in the PRC products (Menasherow et al. 2016). Alternative method to differentiate the vaccine from the field strain of LSDV was also developed and used in Greece during the LSD outbreak (Agianniotaki et al. 2017). Two real-time PCR assays for have been developed for the specific detection of field Balkan strains of LSDV (Vidanovic et al. 2016).

The molecular tests described above use genes which are able to characterize capripoxvirus work reasonably well. Since capripoxviruses are double-stranded DNA viruses, they are genetically more stable compared to single-stranded DNA or RNA viruses. However, since poxviruses can recombine with each other, it is possible that in a rare event in which recombination occurs (Gershon et al. 1989) that the assays described above will not be able to correctly classify the virus. To ensure the proper identity of the capripoxvirus species, full-length genomic sequencing is the most appropriate method. However due to the cost of full-length sequencing, it is not routinely performed. For this reason molecular epidemiology for capripoxviruses is not as advanced compared to many other viruses of veterinary importance.

Serology has been developed to characterize the immune response following infection with LSDV. The gold standard and first developed serological test was the virus neutralizing test which measures the neutralizing antibodies against the virus. The development of additional serological assays for the detection of capripoxvirus-specific antibodies has been challenging. Immunostaining (Babiuk et al. 2007) and indirect florescence antibody testing (IFAT) (Gari et al. 2008, 2012) have been developed as additional serological methods. However, these tests are labour intensive, time-consuming and require working with a live virus. An indirect ELISA based on P32 as an antigen expressed in *E. coli* was developed by Carn et al. (1994) and further evaluated by Heine et al. (1999); unfortunately there was limited sera tested on using the assay, and there are difficulties in producing the P32 protein due to stability issues. The P32 antigen has been further developed as a synthetic peptide ELISA and evaluated using sera from sheep and goats infected with sheeppox and goatpox (Tian et al. 2010). Unfortunately this assay has not been evaluated with sera from LSD infected cattle. An indirect ELISA based on whole heat-inactivated sheeppox virus has been described (Babiuk et al. 2009). This whole virus ELISA was evaluated with 276 cattle sera samples and had a diagnostic sensitivity and specificity of 88% and 97%, respectively. Although this ELISA is not available due to the expense and difficulties in producing the antigen, it demonstrated that it is feasible to develop an ELISA for capripoxviruses. Following this study, 42 different proteins were evaluated from various open reading frames of the capripoxvirus genome (Bowden et al. 2009). The screening of these antigens revealed two viral core proteins that were developed into an indirect ELISA. Although initial evaluation of this ELISA appeared promising, further testing revealed that the viral core antigen ELISA was only able to detect experimentally infected cattle sera but not LSD VNT positive cattle sera from the field. The likely reason for this is that the VNT detects capripoxvirus surface proteins, whereas the core ELISA detects antibodies to internal viral core proteins. With further validation, this assay could potentially be used to detect recent infections; however, it is not suitable for classical surveillance purposes. The development of capripoxvirus ELISA is complicated as there are many antigens, and the immunogenicity of these antigens has not been fully characterized. A mass screening approach using every capripoxvirus protein would be able to identify antigens suitable for an ELISA.

It is critical that diagnostic assays will be validated according to OIE standards and that diagnostic laboratories co-ordinate the validation of tests using the same assay protocols and reagents. This is especially important for serology as the virus neutralization test should be standardized between laboratories with the cells and virus used for neutralization as well as the sera dilutions used. The lack of an ELISA has severely hampered serosurveillance for capripoxvirus. In 2017 the first ELISA detecting antibodies against LSDV become commercially available.

Histology can be used for diagnosis of LSD. In skin lesions, there are A-type inclusion bodies and characteristic so-called sheeppox cells which indicate of lumpy skin disease. Immunohistochemistry using either polyclonal capripoxvirus antisera or a monoclonal antibody specific for capripoxvirus can be used to demonstrate capripoxvirus-specific antigen in tissues (Awadin et al. 2011).

References

Agianniotaki EI, Tasioudi KE, Chaintoutis SC, Iliadou P, Mangana-Vougiouka O, Kirtzalidou A, Alexandropoulos T, Sachpatzidis A, Plevraki E, Dovas CI, Chondrokouki E (2017) Lumpy skin disease outbreaks in Greece during 2015–16, implementation of emergency immunization and genetic differentiation between field isolates and vaccine virus strains. Vet Microbiol 201:78–84

Awadin W, Hussein H, Elseady Y, Babiuk S, Furuoka H (2011) Detection of lumpy skin disease virus antigen and genomic DNA in formalin-fixed paraffin-embedded tissues from an Egyptian outbreak in 2006. Transbound Emerg Dis 58:451–457

Babiuk S, Parkyn G, Copps J, Larence JE, Sabara MI, Bowden TR, Boyle DB, Kitching RP (2007) Evaluation of an ovine testis cell line (OA3.Ts) for propagation of capripoxvirus isolates and development of an immunostaining technique for viral plaque visualization. J Vet Diagn Invest 19:486–491

Babiuk S, Wallace DB, Smith SJ, Bowden TR, Dalman B, Parkyn G, Copps J, Boyle DB (2009) Detection of antibodies against capripoxviruses using an inactivated sheeppox virus ELISA. Transbound Emerg Dis 56:132–141

Balinsky CA, Delhon G, Smoliga G, Prarat M, French RA, Geary SJ, Rock DL, Rodriguez LL (2008) Rapid preclinical detection of sheeppox virus by a real-time PCR assay. J Clin Microbiol 46:438–442

Bowden TR, Babiuk SL, Parkyn GR, Copps JS, Boyle DB (2008) Capripoxvirus tissue tropism and shedding: a quantitative study in experimentally infected sheep and goats. Virology 371: 380–393

Bowden TR, Coupar BE, Babiuk SL, White JR, Boyd V, Duch CJ, Shiell BJ, Ueda N, Parkyn GR, Copps JS, Boyle DB (2009) Detection of antibodies specific for sheeppox and goatpox viruses using recombinant capripoxvirus antigens in an indirect enzyme-linked immunosorbent assay. J Virol Methods 161:19–29

Carn VM, Kitching RP, Hammond JM, Chand P (1994) Use of a recombinant antigen in an indirect ELISA for detecting bovine antibody to capripoxvirus. J Virol Methods 49:285–294

Das A, Babiuk S, McIntosh MT (2012) Development of a loop-mediated isothermal amplification assay for rapid detection of capripoxviruses. J Clin Microbiol 50:1613–1620

Gari G, Biteau-Coroller F, LeGoff C, Caufour P, Roger F (2008) Evaluation of indirect fluorescent antibody test (IFAT) for the diagnosis and screening of lumpy skin disease using Bayesian method. Vet Microbiol 129:269–280

Gari G, Grosbois V, Waret-Szkuta A, Babiuk S, Jacquiet P, Roger F (2012) Lumpy skin disease in Ethiopia: seroprevalence study across different agro-climate zones. Acta Trop 123:101–106

Gelaye E, Lamien CE, Silber R, Tuppurainen ES, Grabherr R, Diallo A (2013) Development of a cost-effective method for capripoxvirus genotyping using snapback primer and dsDNA intercalating dye. PLoS One 8:e75971

Gelaye E, Mach L, Kolodziejek J, Grabherr R, Loitsch A, Achenbach JE, Nowotny N, Diallo A, Lamien CE (2017) A novel HRM assay for the simultaneous detection and differentiation of eight poxviruses of medical and veterinary importance. Sci Rep 7:42892

Gershon PD, Kitching RP, Hammond JM, Black DN (1989) Poxvirus genetic recombination during natural virus transmission. J Gen Virol 70:485–489

Gulbahar MY, Davis WC, Yuksel H, Cabalar M (2006) Immunohistochemical evaluation of inflammatory infiltrate in the skin and lung of lambs naturally infected with sheeppox virus. Vet Pathol 43:67–75

Heine HG, Stevens MP, Foord AJ, Boyle DB (1999) A capripoxvirus detection PCR and antibody ELISA based on the major antigen P32, the homolog of the vaccinia virus H3L gene. J Immunol Methods 227:187–196

Ireland DC, Binepal YS (1998) Improved detection of capripoxvirus in biopsy samples by PCR. J Virol Methods 74:1–7

Kitching RP (1986) Passive protection of sheep against capripoxvirus. Res Vet Sci 41:247–250

Kitching RP, Smale C (1986) Comparison of the external dimensions of capripoxvirus isolates. Res Vet Sci 41:425–427

Lamien CE, Le Goff C, Silber R, Wallace DB, Gulyaz V, Tuppurainen E, Madani H, Caufour P, Adam T, El Harrak M, Luckins AG, Albina E, Diallo A (2011a) Use of the Capripoxvirus homologue of Vaccinia virus 30 kDa RNA polymerase subunit (RPO30) gene as a novel diagnostic and genotyping target: development of a classical PCR method to differentiate Goat poxvirus from Sheep poxvirus. Vet Microbiol 149:30–39

Lamien CE, Lelenta M, Goger W, Silber R, Tuppurainen E, Matijevic M, Luckins AG, Diallo A (2011b) Real time PCR method for simultaneous detection, quantitation and differentiation of capripoxviruses. J Virol Methods 171:134–140

Le Goff C, Lamien CE, Fakhfakh E, Chadeyras A, Aba-Adulugba E, Libeau G, Tuppurainen E, Wallace DB, Adam T, Silber R, Gulyaz V, Madani H, Caufour P, Hammami S, Diallo A, Albina E (2009) Capripoxvirus G-protein-coupled chemokine receptor: a host-range gene suitable for virus animal origin discrimination. J Gen Virol 90:1967–1977

Mangana-Vougiouka O, Markoulatos P, Koptopoulos G, Nomikou K, Bakandritsos N, Papadopoulos O (1999) Sheep poxvirus identification by PCR in cell cultures. J Virol Methods 77:75–79

Menasherow S, Rubinstein-Giuni M, Kovtunenko A, Eyngor Y, Fridgut O, Rotenberg D, Khinich Y, Stram Y (2014) Development of an assay to differentiate between virulent and vaccinestrains of lumpy skin disease virus (LSDV). J Virol Methods 199:95–101

Menasherow S, Erster O, Rubinstein-Giuni M, Kovtunenko A, Eyngor E, Gelman B, Khinich E, Stram Y (2016) A high-resolution melting (HRM) assay for the differentiation between Israeli field and Neethling vaccine lumpy skin disease viruses. J Virol Methods 232:12–15

Murray L, Edwards L, Tuppurainen ES, Bachanek-Bankowska K, Oura CA, Mioulet V, King DP (2013) Detection of capripoxvirus DNA using a novel loop-mediated isothermal amplification assay. BMC Vet Res 9:90

Orlova ES, Shcherbakova AV, Diev VI, Zakharov VM (2006) Differentiation of capripoxvirus species and strains by polymerase chain reaction. Mol Biol (Mosk) 40:158–164

Roeder P (2011) Making a global impact: the eradication of rinderpest. Vet Rec 169:650–652

Shalaby MA, El-Deeb A, El-Tholoth M, Hoffmann D, Czerny CP, Hufert FT, Weidmann M, Abd El Wahed A (2016) Recombinase polymerase amplification assay for rapid detection of lumpy skin disease virus. BMC Vet Res 12:244

Stubbs S, Oura CA, Henstock M, Bowden TR, King DP, Tuppurainen ES (2012) Validation of a high-throughput real-time polymerase chain reaction assay for the detection of capripoxviral DNA. J Virol Methods 179:419–422

Tian H, Chen Y, Wu J, Shang Y, Liu X (2010) Serodiagnosis of sheeppox and goatpox using an indirect ELISA based on synthetic peptide targeting for the major antigen P32. Virol J 7:245

Tuppurainen ES, Pearson CR, Bachanek-Bankowska K, Knowles NJ, Amareen S, Frost L, Henstock MR, Lamien CE, Diallo A, Mertens PP (2014) Characterization of sheep pox virus vaccine for cattle against lumpy skin disease virus. Antivir Res 109:1–6

Vidanovic D, Sekler M, Petrovic T, Debeljak Z, Vaskovic N, Matovic K, Hoffmann B (2016) Real-time PCR assays for the specific detection of field Balkan strains of lumpy skin disease virus. Acta Vet Beograd 66:444–454

Zhao Z, Fan B, Wu G, Yan X, Li Y, Zhou X, Yue H, Dai X, Zhu H, Tian B, Li J, Zhang Q (2014) Development of loop-mediated isothermal amplification assay for specific and rapid detection of differential goat pox virus and sheep pox virus. BMC Microbiol 14:10

Zheng M, Liu Q, Jin N, Guo J, Huang X, Li H, Zhu W, Xiong Y (2007) A duplex PCR assay for simultaneous detection and differentiation of Capripoxvirus and Orf virus. Mol Cell Probes 21:276–281

Treatment of Lumpy Skin Disease

17

Shawn Babiuk

Unfortunately there are no specific antiviral drugs available for the treatment of lumpy skin disease. The only treatment available is supportive care of cattle. This can include treatment of skin lesions using wound care sprays and the use of antibiotics to prevent secondary skin infections and pneumonia. Anti-inflammatory painkillers can be used to keep up the appetite of affected animals. Intravenous fluid administration may be of benefit; however this may not be practical in the field. The lack of treatment options for lumpy skin disease virus emphasizes the need of using effective vaccination for preventing disease.

© Springer International Publishing AG, part of Springer Nature 2018
E. S. M. Tuppurainen et al., *Lumpy Skin Disease*,
https://doi.org/10.1007/978-3-319-92411-3_17

Part III

Control and Eradication

Vaccines Against LSD and Vaccination Strategies

18

Shawn Babiuk

Since there are no drug treatments for lumpy skin disease (LSD), vaccination using live attenuated vaccines are used to control the disease. The main prerequisites of a good vaccine are safety and protection from infection. The safety of a vaccine is determined by the frequency and severity of the adverse reactions it might cause, by the probability of reversion to virulence and by its purity. Protection provided by a vaccine depends on the specific immunologic response it elicits and can be measured by its efficacy or effectiveness. The efficacy of a vaccine is represented by the percentage of morbidity prevented by vaccination. Since capripoxviruses are genetically similar and have no serotypes, it was previously suggested that the development of a single capripoxvirus vaccine to protect against all three species should be possible (Kitching 1983, 2003). However, currently there is no universal vaccine available. There are several reasons for this. One reason is that although capripoxvirus vaccines may be effective for a particular host, they may not be as effective in a different host due to being not fully attenuated, over attenuated and/or not immunogenic. This is due to the complex virus-host interactions which determine virulence as well as the immune response elicited. The quality control and production of the vaccines may not follow the guidelines provided by the good manufacturing practices (GMP) for biological products by the World Health Organization. An additional reason is the geographic distributions of sheeppox, goatpox and LSD, which are partially different. There are countries which have only sheep- and/or goatpox without LSD, countries which only have LSD and countries which have sheep- and/or goatpox as well as LSD (Babiuk et al. 2008). Due to potential safety issues, affected countries will tend to use only those capripoxvirus vaccines containing viruses present in the country (Tuppurainen and Oura 2012). This has led to regulatory challenges to authorize the use of these vaccines in the country or region.

Different vaccines have been generated from sheeppox, goatpox and LSD viruses. These live attenuated capripoxvirus vaccines have been produced by growing capripoxviruses of different origins several times on a wide variety of different cells (Davies and Mbugwa 1985) and/or embryonated chicken eggs (van Rooyen

© Springer International Publishing AG, part of Springer Nature 2018
E. S. M. Tuppurainen et al., *Lumpy Skin Disease*,
https://doi.org/10.1007/978-3-319-92411-3_18

et al. 1969). The level of attenuation and ability to provide immunity in different hosts such as sheep, goats and cattle are not always fully characterized. Some vaccines might be substantially attenuated in one host but too virulent to be used in another. In addition, some of the vaccines do not have the full-genome sequence available, and the genetic changes responsible for attenuation are not fully characterized. Despite this, these vaccines are used throughout endemic regions to protect against sheeppox, goatpox and/or LSD virus.

Currently LSD vaccines which have been derived from a lumpy skin disease virus (LSDV) are considered the most effective. These vaccines were generated by multiple passages on cells and some passaging additionally on eggs. Sequencing of the full genome has been carried out for the attenuated LSDV field strain (SIS) Lumpyvax® (MSD Animal Health), which is 150,480 bp, and for Lumpy Skin Disease Vaccine for Cattle (Onderstepoort Biological Products—OBP) which is 150,508 bp of length. These vaccines share 99.9% homology with minor differences (Mathijs et al. 2016). These LSDV-derived vaccines have been the most extensively evaluated and have been demonstrated to be effective in cattle. Live attenuated LSD vaccines may cause a local granulomatous skin reaction at the site of inoculation. In addition, fever and the reduction in milk production can follow vaccination. Under certain circumstances, these vaccines can cause a low incidence of Neethling-associated disease (Ben-Gera et al. 2015). In a study performed in 2013 during a large LSD epidemic, which took place in Israel, a commercial LSDV-based vaccine was compared to a sheeppox (RM-65) (Jovivac®, Jordan) live attenuated vaccine administered at the same dose. Both vaccines were randomly and simultaneously administered to 4694 cows in 15 dairy herds. The effectiveness of the LSD vaccine as compared to the sheeppox vaccine was 77% for preventing a laboratory-confirmed disease. No data could be recovered from this study on the effectiveness of the sheeppox vaccine as there were no placebo controls in this study. Only 0.38% of the cows vaccinated with the attenuated LSD vaccine developed a generalized adverse reaction caused by the vaccine virus. Those animals showed a mild skin reaction characterized by appearance of small lumps over the body within the first 2 weeks after vaccination (Ben-Gera et al. 2015). The findings of this study were later re-enforced by an analysis of data from an outbreak in Greece during 2016. As a response to LSD outbreaks, vaccination in the region of Serres in Greece was performed along with the advancement of the outbreak. This allowed the comparison between the morbidity in vaccinated and non-vaccinated herds. The analysis revealed that 2 weeks and 1 month after vaccination, vaccine effectiveness reached 62.5% and 80%, respectively (AHAW 2015). Another study evaluated adverse reactions following immunization of cattle with the OBP vaccine and revealed 12% (26/215) of vaccinated cattle developed swelling at the injection site and that 9% of vaccinated animals developed small cutaneous lumps between 8 and 18 days following vaccination (Katsoulos et al. 2017). The Ethiopian LSD Neethling vaccine failed to provide protection in cattle against wild Ethiopian strain of LSDV either no adverse effects were observed in the vaccinated calves (Gari et al. 2015). The likely reason for the failure of the Ethiopian Neethling vaccine was due to issues with the vaccine seed stock. This illustrates well the critical importance of having a quality vaccine production system.

There are several sheeppox and goatpox vaccines which have been used in cattle against LSDV. The Kenyan sheep- and goatpox vaccines derived from O-240 as KS-1 36 3/95, KSGP O-240 and Kenyavac KSGP O-240 were isolated from sheep and characterized as LSDV using gene sequencing of the RPO30 and GPCR genes (Davies 1976; Tuppurainen et al. 2014). Another isolate KSGP O-180 was isolated from the same outbreak as the KSGP O-240 and is also likely to be a LSDV. The KSGP O-180 had been attenuated by passaging the virus 18 times on bovine fetal muscle cells (Davies and Mbugwa 1985), and the KSGP O-240 was attenuated after only 6 passages on cell cultures (Kitching et al. 1987). The Kenyan sheep- and goatpox virus vaccines O-240 and O-180 have been used successfully as a vaccine for sheeppox and goatpox (Davies and Mbugwa 1985; Kitching et al. 1987). However, when used in cattle, due to a low level of attenuation, these vaccines can cause clinical signs of LSD in vaccinated cattle. There are several studies which identify problems using these vaccines in cattle. Also Israelis reported generalized reactions, typical of LSD, occurring in dairy cattle vaccinated with an attenuated Kenya sheep- and goatpox strain 0240 (Yeruham et al. 1994). Adverse reactions were also observed following LSD vaccination campaign in Jordan, despite a lack of proper record keeping of the two different vaccines used to identify which vaccine caused the adverse reactions (Abutarbush et al. 2016). These adverse reactions included fever, decreased milk production as well as skin nodules. The vaccines used were the RM65 isolate and an unlabelled vaccine which was identified as a LSDV isolate (likely a Kenyan sheep- and goatpox vaccine) by PCR. The adverse reactions seen after the use of the sheeppox RM65 vaccine were much milder than those seen after administration of the LSDV vaccine (Abutarbush et al. 2016). During the 2006 outbreak of LSD in Egypt, it was reported that the live attenuated KSGP O-240 strain did not provide cattle with complete protection against LSD (Salib and Osman 2011). In Ethiopia there were reported morbidity and mortality rates of 22.9% and 2.31%, respectively, in fully vaccinated herds with the Kenyan sheeppox vaccine (Ayelet et al. 2013). In 2008 and 2009, reinfection of vaccinated animals was observed during LSDV epidemics (Ayelet et al. 2014). In Ethiopia, the low performance of the local Kenyan sheep- and goatpox vaccine, together with insufficient vaccination coverage, allows LSD outbreaks to occur (Gelaye et al. 2015). In Oman vaccination of cattle with the Kenyan sheep- and goatpox vaccine O-240 was not successful (Somasundaram 2011). With these studies it should be taken into account that the vaccination coverage in Ethopia and Oman is very poor as well as quality control issue with vaccine production, which may be reasons for the vaccine failure observed in these studies. The effectiveness of the Kenyan sheep- and goatpox virus vaccines is not ideal. Considering the adverse vaccine reactions likely to be caused by under-attenuated Kenyan O-240 and O-180 in cattle these Kenyan vaccines should not be used in cattle.

Several studies have shown the efficacy of goatpox viruses for protecting cattle from challenge by LSDV. The Kedong and Isiolo viruses were isolated from sheep in Kenya during the 1950s but were later shown to actually be goatpox viruses

(Tuppurainen et al. 2014). Both were shown to protect cattle from LSDV challenge (Coakley and Capstick 1961). In a recent study in Ethiopia, a commercially available attenuated Gorgon goatpox strain vaccine was administered at two doses to ten calves (five calves in each dose group). This study compared the efficacy and immunogenicity of the National Veterinary Institute (NVI) LSDV Neethling, KSGP O-180 and the Gorgan GTP vaccine (Caprivac™, Jordan Bio-Industries Center, Amman, Jordan). The study included vaccine challenge experiments in a controlled environment and monitoring of immune responses in vaccinated animals in the field. The Ethiopian LSDV Neethling and KSGP O-180 vaccines failed to provide protection in cattle against LSDV, whereas the Gorgan goatpox vaccine protected all the vaccinated calves from clinical signs of LSD. All calves were completely protected from challenge by a wild Ethiopian strain of LSDV. Moreover, the Gorgan goatpox-vaccinated cattle showed higher levels of cellular immune responses at the vaccination site, consistent with greater immunogenicity (Gari et al. 2015). No adverse effects were observed in the vaccinated calves (Gari et al. 2015). Attenuated goatpox vaccines currently seem like a good alternative to vaccination by homologous vaccines. However, there is a need for provision of more extensive data from the field and for performance of larger clinical trials to ensure the efficacy and safety of these vaccines.

There have been different sheeppox vaccines used in cattle against LSD. The Yugoslavian RM65 SPPV vaccine, at a ten times higher dose than indicated for sheep, has commonly been used for cattle across the Middle East. Incomplete protection was observed when the RM65 SPP vaccine was used to vaccinate cattle against LSDV in Israel at the same dose as recommended for sheep from 2006 to 2007 (Brenner et al. 2009). A later study in Israel compared both LSDV and RM65 (10X) vaccines in cattle. Both vaccines were safe to use, although mild adverse effects were observed after vaccination using the LSDV vaccine. However, the efficacy of LSDV vaccine was superior when compared with the RM65 SPPV (10X) vaccine (Ben-Gera et al. 2015). There was no evidence for protection elicited by the RM65 vaccine as this was the reference group, and a high incidence was observed in fully vaccinated herds. Moreover, the Neethling vaccine had an effectiveness of 75% compared to this vaccine, which is very similar to the effectiveness calculated for this vaccine in Greece. The Romanian strain sheeppox vaccine has been used in cattle in Egypt and appeared to be immunogenic (Davies 1991). In Turkey, the Bakirköy SPPV at three to four times the recommended dose for sheep has been used in cattle against LSD virus. No statistically significant differences in disease incidence were demonstrated between cattle vaccinated with this vaccine and non-vaccinated cattle (Şevik and Doğan 2017). Though the vaccine was extensively used in Turkey, the disease has persisted in the country for several years and still exists there. In Russia, the local sheeppox vaccine is currently being used to vaccinate cattle against LSD. For the successful control of LSD, it is essential that prior to the use of a vaccine, the efficacy and safety of the vaccine candidate are evaluated in controlled challenge studies.

Vaccine failure can occur due to a number of reasons, most common being vaccinating in the presence of an outbreak. The other reasons include vaccine manufacturing issues, reusing of needles, inappropriate administration of the vaccine, storage and breakdown in the cold chain and the use of illegal vaccines. Reports illustrating vaccine failure highlight the need for improved testing of vaccines as well as for monitoring vaccines for efficacy as well as adverse reactions. The field experience obtained from the most recent outbreaks of LSDV in the Middle East and the Horn of Africa indicates that cross protection provided by non-homologous vaccine viruses is only partial.

Generally live attenuated vaccines generate more broad protective immunity compared to non-replicating vaccines. Inactivated LSD vaccines have not been commercially available, but recently a novel killed vaccine for cattle against LSD has been developed, and field experiments are ongoing. The duration of immunity elicited by a killed LSD vaccine is not as long as for the live attenuated vaccines, and a second immunization is required in order to generate protective immunity comparable to a single vaccination with a live attenuated vaccine (Boumart et al. 2016). A killed LSD vaccine could be useful for export/import of live cattle purposes.

Vaccinia virus (VV) was one of the first viruses to be demonstrated as a vaccine vector. This demonstration has led to the development of various poxviruses to be used as vaccine vectors. The utility of having multiple different poxvirus vector platforms available allows for the development of many different vectored vaccines and the flexibility to use different vector platforms in different target animals. LSDV has experimentally been demonstrated as an effective vaccine vector for many different antigens in sheep, goats as well as cattle (Boshra et al. 2013). The KS-1 vaccine has been demonstrated to protect cattle against rinderpest using either the hemagglutinin gene (Romero et al. 1994a) or the fusion gene (Romero et al. 1994b) inserted into the thymidine kinase (TK) region. In addition the fusion gene construct was demonstrated to protect goats against peste des petits ruminants (PPR) virus (Romero et al. 1995). The KS-1 vaccine with either the F gene (Diallo et al. 2002) or the H gene (Berhe et al. 2003) from PPR virus was able to protect goats against challenge. The bluetongue antigens VP7 (Wade-Evans et al. 1996), VP2, NS1 and NS3 (Perrin et al. 2007) in the KS-1 vaccine were able to partially protect sheep and goats. The glycoproteins Gn and Gc from Rift Valley fever virus in the KS-1 vaccine have been able to generate Rift Valley fever neutralizing antibodies in sheep (Wallace et al. 2006; Soi et al. 2010).

Due to the many non-essential genes present in LSDV which are involved in the modulation of the host immune system, it is possible to generate gene-deleted or gene-disrupted viruses which may be attenuated. Despite the numerous potential different gene targets that could be used to generate attenuated LSD viruses, there are limited studies which have only evaluated a small number of gene targets. The thymidine kinase gene has been used to generate recombinant experimental LSDV vaccines (Wallace and Viljoen 2005). In sheeppox the Kelch-like gene SPPV-019 has been demonstrated to attenuate a virulent sheeppox virus (Balinsky et al. 2007). It has been demonstrated that a LSDV with the IL-10 gene disrupted is safe in sheep and goats and can protect against sheeppox and goatpox although this virus is only partially attenuated in cattle (Boshra et al. 2015). It is very likely that LSDV can be attenuated in several different ways through gene deletion of various non-essential genes and combinations of these genes. Furthermore, it is likely that targeted molecular modification of the existing LSDV vaccines could be used to enhance the immunogenicity of the vaccines as well as decrease the injection site reactions caused by these vaccines.

Unfortunately there is no current way of differentiating infected from vaccinated animals (DIVA). The principle behind a DIVA vaccine is that a gene which generates an antibody response is deleted from the vaccine making a negative marker vaccine (van Oirschot et al. 1986; Pasick 2004; Vannie et al. 2007). Animals vaccinated with a DIVA vaccine will not develop antibodies to the deleted gene which are detected by the companion serological assay to the specific deleted gene. If vaccinated animals are then infected, they will generate antibodies to the companion serological assay as the natural virus will contain the antigen. The need for such vaccine is illustrated in dealing with an outbreak as it is impossible to implement an effective serosurveillance programme when vaccination is used to control LSD. The lack of a DIVA vaccine also makes it difficult to determine when to stop vaccinating to gain freedom of disease status. Although making a DIVA vaccine is straightforward for LSD in respect to generating the attenuated virus, the major issue is there is no validated ELISA test to identify the antigen to delete.

18.1   Vaccination Strategies

With sheeppox and goatpox, there is a precedent that these diseases can be eradicated using stamping out, animal movement restrictions and quarantine. This was demonstrated in England in 1847 (Simmons 1874). Unfortunately, with LSD it is extremely difficult to control using only total or modified stamping out and animal movement controls along with quarantine. Israel was able to regain freedom from disease following the first outbreaks which occurred in 1989 through slaughter of all the sheep, goats and cattle in the affected area (Yeruham et al. 1995). It also occurred after the 2006 and 2007 outbreaks. The RM-65 sheep dose vaccine was used there as well, but the evidence gathered to date suggests it is very ineffective. This demonstrated that modified stamping out actually controlled the disease, likely due to low cattle density in the affected regions.

Mathematical models have been used to evaluate different control measure responses to the spread of LSDV outbreaks by the European Food Safety Authority (EFSA). This study revealed that vaccination had the greater impact in reducing LSDV spread compared to any culling policy, even when low vaccination effectiveness is considered (EFSA Panel on Animal Health and Welfare 2016).

Today there is no country where LSD outbreaks have occurred that is considered free of disease by the OIE definition. Due to mass vaccination, Israel currently does not have any clinical LSD. However, as the disease is endemic in the region and in some areas sufficient vaccination coverage have not been achieved, due to the vector-borne spread of LSDV, outbreaks are likely to reoccur in case the vaccinations are stopped.

The experiences obtained from Israel and the Balkan region demonstrate that the spread of LSD can successfully be contained using a well-organized vaccination campaign with sufficient coverage and effective vaccines. The LSD outbreak situation in South Africa where there are frequent LSD outbreaks is likely due to the low vaccine coverage, and vaccination is practised regularly in large commercial cattle farms whereas more rarely by small holders and backyard farmers.

Since November 2016, the European Commission (EC) allows preventive vaccination if the EC has been notified with the vaccination plan. In addition, a status "freedom-with-vaccination" is accepted, allowing countries to apply preventative vaccination in order to create vaccinated zones to stop the spread of LSDV from affected neighbouring countries. The size of the vaccination zone needs to be determined as the distance LSDV can spread is not fully understood. It has been suggested that LSD spread by long-distance dispersal of vectors by winds from outbreaks in Egypt from Damietta and Port Said to Israel (Klausner et al. 2015). More importantly, the long-distance spread is likely to occur via legal or illegal movement of live animals from affected countries. In currently affected regions, internationally coordinated, harmonized vaccination campaign is the cornerstone of the eradication of LSD and regaining the disease-free status.

References

Abutarbush SM, Hananeh WM, Ramadan W, Al Sheyab OM, Alnajjar AR, Al Zoubi IG, Knowles NJ, Bachanek-Bankowska K, Tuppurainen ES (2016) Adverse reactions to field vaccination against lumpy skin disease in Jordan. Transbound Emerg Dis 63:e213–e219

AHAW (2015) Scientific opinion on lumpy skin disease. EFSA J 13:3986

Ayelet G, Abate Y, Sisay T, Nigussie H, Gelaye E, Jemberie S, Asmare K (2013) Lumpy skin disease: preliminary vaccine efficacy assessment and overview on outbreak impact in dairy cattle at Debre Zeit, central Ethiopia. Antivir Res 98:261–265

Ayelet G, Haftu R, Jemberie S, Belay A, Gelaye E, Sibhat B, Skjerve E, Asmare K (2014) Lumpy skin disease in cattle in central Ethiopia: outbreak investigation and isolation and molecular detection of the virus. Rev Sci Tech 33:877–887

Babiuk S, Bowden TR, Boyle DB, Wallace DB, Kitching RP (2008) Capripoxviruses: an emerging worldwide threat to sheep, goats and cattle. Transbound Emerg Dis 55:263–272

Balinsky CA, Delhon G, Afonso CL, Risatti GR, Borca MV, French RA, Tulman ER, Geary SJ, Rock DL (2007) Sheeppox virus kelch-like gene SPPV-019 affects virus virulence. J Virol 81:11392–11401

Ben-Gera J, Klement E, Khinich E, Stram Y, Shpigel NY (2015) Comparison of the efficacy of Neethling lumpy skin disease virus and x10RM65 sheep-pox live attenuated vaccines for the prevention of lumpy skin disease: the results of a randomized controlled field study. Vaccine 33:4837–4842

Berhe G, Minet C, Le Goff C, Barrett T, Ngangnou A, Grillet C, Libeau G, Fleming M, Black DN, Diallo A (2003) Development of a dual recombinant vaccine to protect small ruminants against peste-des-petits-ruminants virus and capripoxvirus infections. J Virol 77:1571–1577

Boshra H, Truong T, Nfon C, Gerdts V, Tikoo S, Babiuk LA, Kara P, Mather A, Wallace D, Babiuk S (2013) Capripoxvirus-vectored vaccines against livestock diseases in Africa. Antivir Res 98: 217–227

Boshra H, Truong T, Nfon C, Bowden TR, Gerdts V, Tikoo S, Babiuk LA, Kara P, Mather A, Wallace DB, Babiuk S (2015) A lumpy skin disease virus deficient of an IL-10 gene homologue provides protective immunity against virulent capripoxvirus challenge in sheep and goats. Antivir Res 123:39–49

Boumart Z, Daouam S, Belkourati I, Rafi L, Tuppurainen E, Tadlaoui KO, El Harrak M (2016) Comparative innocuity and efficacy of live and inactivated sheeppox vaccines. BMC Vet Res 12:133

Brenner J, Bellaiche M, Gross E, Elad D, Oved Z, Haimovitz M, Wasserman A, Friedgut O, Stram Y, Bumbarov V, Yadin H (2009) Appearance of skin lesions in cattle populations vaccinated against lumpy skin disease: statutory challenge. Vaccine 27:1500–1503

Coakley W, Capstick PB (1961) Protection of cattle against lumpy skin disease. Res Vet Sci 12: 123–127

Davies FG (1976) Characteristics of a virus causing a pox disease in sheep and goats in Kenya, with observation on the epidemiology and control. J Hyg (Lond) 76:163–171

Davies FG (1991) Lumpy skin disease of cattle: a growing problem in Africa and the Near East. World Anim Rev 68:37–42

Davies FG, Mbugwa G (1985) The alterations in pathogenicity and immunogenicity of a Kenya sheep and goat pox virus on serial passage in bovine foetal muscle cell cultures. J Comp Pathol 95:565–572

Diallo A, Minet C, Berhe G, Le Goff C, Black DN, Fleming M, Barrett T, Grillet C, Libeau G (2002) Goat immune response to capripox vaccine expressing the hemagglutinin protein of peste des petits ruminants. Ann N Y Acad Sci 969:88–91

EFSA Panel on Animal Health and Welfare (2016) Urgent advice on lumpy skin disease. EFSA J 14:e04573

Gari G, Abie G, Gizaw D, Wubete A, Kidane M, Asgedom H, Bayissa B, Ayelet G, Oura CA, Roger F, Tuppurainen ES (2015) Evaluation of the safety, immunogenicity and efficacy of three capripoxvirus vaccine strains against lumpy skin disease virus. Vaccine 33:3256–3261

Gelaye E, Belay A, Ayelet G, Jenberie S, Yami M, Loitsch A, Tuppurainen E, Grabherr R, Diallo A, Lamien CE (2015) Capripox disease in Ethiopia: genetic differences between field isolates and vaccine strain, and implications for vaccination failure. Antivir Res 119:28–35

Katsoulos PD, Chaintoutis SC, Dovas CI, Polizopoulou ZS, Brellou GD, Agianniotaki EI, Tasioudi KE, Chondrokouki E, Papadopoulos O, Karatzias H, Boscos C (2017) Investigation on the incidence of adverse reactions, viraemia and haematological changes following field immunization of cattle using a live attenuated vaccine against lumpy skin disease. Transbound Emerg Dis. https://doi.org/10.1111/tbed.12646

Kitching P (1983) Progress towards sheep and goat pox vaccines. Vaccine 1:4–9

Kitching RP (2003) Vaccines for lumpy skin disease, sheep pox and goat pox. Dev Biol (Basel) 114:161–167

Kitching RP, Hammond JM, Taylor WP (1987) A single vaccine for the control of capripox infection in sheep and goats. Res Vet Sci 42:53–60

Klausner Z, Fattal E, Klement E (2015) Using synoptic systems' typical wind trajectories for the analysis of potential atmospheric long-distance dispersal of lumpy skin disease virus. Transbound Emerg Dis 64:398–410

Mathijs E, Vandenbussche F, Haegeman A, King A, Nthangeni B, Potgieter C, Maartens L, Van Borm S, De Clercq K (2016) Complete genome sequences of the Neethling-like lumpy skin disease virus strains obtained directly from three commercial live attenuated vaccines. Genome Announc 4:e01255-16

Pasick J (2004) Application of DIVA vaccines and their companion diagnostic tests to foreign animal disease eradication. Anim Health Res Rev 5:257–262

Perrin A, Albina E, Bréard E, Sailleau C, Promé S, Grillet C, Kwiatek O, Russo P, Thiéry R, Zientara S, Cêtre-Sossah C (2007) Recombinant capripoxviruses expressing proteins of bluetongue virus: evaluation of immune responses and protection in small ruminants. Vaccine 25: 6774–6783

Romero CH, Barrett T, Chamberlain RW, Kitching RP, Fleming M, Black DN (1994a) Recombinant capripoxvirus expressing the hemagglutinin protein gene of rinderpest virus: protection of cattle against rinderpest and lumpy skin disease viruses. Virology 204:425–429

Romero CH, Barrett T, Kitching RP, Carn VM, Black DN (1994b) Protection of cattle against rinderpest and lumpy skin disease with a recombinant capripoxvirus expressing the fusion protein gene of rinderpest virus. Vet Rec 135:152–154

Romero CH, Barrett T, Kitching RP, Bostock C, Black DN (1995) Protection of goats against peste des petits ruminants with recombinant capripoxviruses expressing the fusion and haemagglutinin protein genes of rinderpest virus. Vaccine 13:36–40

Salib FA, Osman AH (2011) Incidence of lumpy skin disease among Egyptian cattle in Giza Governorate, Egypt. Vet World 4:162–167

Şevik M, Doğan M (2017) Epidemiological and molecular studies on lumpy skin disease outbreaks in Turkey during 2014–2015. Transbound Emerg Dis 64:1268–1279. https://doi.org/10.1111/tbed.12501

Simmons JB (1874) Infectious diseases of animals. J R Agric Soc Engl 10(2):237–240

Soi RK, Rurangirwa FR, McGuire TC, Rwambo PM, DeMartini JC, Crawford TB (2010) Protection of sheep against Rift Valley fever virus and sheep poxvirus with a recombinant capripoxvirus vaccine. Clin Vaccine Immunol 17:1842–1849

Somasundaram MK (2011) An outbreak of lumpy skin disease in a Holstein Dairy Herd in Oman: a clinical report. Asian J Anim Vet Adv 6:851–859

Tuppurainen ES, Oura CA (2012) Review: lumpy skin disease: an emerging threat to Europe, the Middle East and Asia. Transbound Emerg Dis 59:40–48

Tuppurainen ES, Pearson CR, Bachanek-Bankowska K, Knowles NJ, Amareen S, Frost L, Henstock MR, Lamien CE, Diallo A, Mertens PP (2014) Characterization of sheep pox virus vaccine for cattle against lumpy skin disease virus. Antivir Res 109:1–6

van Oirschot JT, Rziha HJ, Moonen PJ, Pol JM, van Zaane D (1986) Differentiation of serum antibodies from pigs vaccinated or infected with Aujeszky's disease virus by a competitive enzyme immunoassay. J Gen Virol 67:1179–1182

van Rooyen PJ, Munz EK, Weiss KE (1969). The optimal conditions for the multiplication of Neethling-type lumpy skin disease virus in embryonated eggs. Onderstepoort J Vet Res 36:165-174.

Vannie P, Capua I, Le Potier MF, Mackay DK, Muylkens B, Parida S, Paton DJ, Thiry E (2007). Marker vaccines and the impact of their use on diagnosis and prophylactic measures. Rev Sci Tech 26:351-372.

Wade-Evans AM, Romero CH, Mellor P, Takamatsu H, Anderson J, Thevasagayam J, Fleming MJ, Mertens PP, Black DN (1996) Expression of the major core structural protein (VP7) of bluetongue virus, by a recombinant capripox virus, provides partial protection of sheep against a virulent heterotypic bluetongue virus challenge. Virology 220:227–231

Wallace DB, Viljoen GJ (2005) Immune responses to recombinants of the South African vaccine strain of lumpy skin disease virus generated by using thymidine kinase gene insertion. Vaccine 23:3061–3067

Wallace DB, Ellis CE, Espach A, Smith SJ, Greyling RR, Viljoen GJ (2006) Protective immune responses induced by different recombinant vaccine regimes to Rift Valley fever. Vaccine 24:7181–7189

Yeruham I, Perl S, Nyska A, Abraham A, Davidson M, Haymovitch M, Zamir O, Grinstein H (1994) Adverse reactions in cattle to a capripox vaccine. Vet Rec 135:330–332

Yeruham I, Nir O, Braverman Y, Davidson M, Grinstein H, Haymovitch M, Zamir O (1995) Spread of lumpy skin disease in Israeli dairy herds. Vet Rec 137:91–93

Slaughter of Infected and In-Contact Animals

19

Eyal Klement

Slaughter of infected and in-contact animals (stamping out) is a policy practised in many countries, mostly for controlling exotic diseases. The epidemiological rational of this action is to prevent continuous virus excretion by infected animals. In some countries, stamping out of LSD is defined by legislation, which is often linked to trade limitations. Stamping out might carry serious economic ramifications that impair people livelihood. In addition, it can severely affect accurate surveillance if not accompanied by adequate and timely compensation, as it may cause severe mistrust between farmers and the veterinary services. Therefore, if not required by legislation, stamping out should be considered carefully by decision-makers before it is implemented. There are several factors that should be taken into account when deciding on stamping out: The epidemic stage on time of detection—stamping out will be much less effective if the epidemic has already spread among many herds. The demographic and political situation in the affected region is of high importance as these may be related to mistrust between the farmers and the government and might have other ramifications. Eventually, the cost of stamping out should be considered against its benefit. These are related to economic impact of LSD as specifically discussed in Chapter 3 of this book. A less radical approach is to perform only partial stamping out, i.e. culling only generalized cases. The rationale behind this policy is based on the assumption that efficient transmission occurs when arthropod vectors bite skin lesions that have significantly higher amounts of virus compared to the amounts of virus in blood, intact skin and other tissues (Babiuk et al. 2008). Performance of partial stamping out necessitates careful in-herd surveillance as its efficiency depends on the early detection of new cases and their fast removal.

The effectiveness of stamping-out strategy for the prevention of LSD outbreak spread is still under debate. During the outbreak in the south of Israel in 1989, all the cattle in the herds in the affected farm were culled (total stamping out). In the outbreaks which occurred in 2006 and 2007, the veterinary services implemented a policy of partial stamping out. All outbreaks were controlled, and the virus was totally eradicated despite the use of a highly non-effective vaccine (AHAW 2015). In

© Springer International Publishing AG, part of Springer Nature 2018

E. S. M. Tuppurainen et al., *Lumpy Skin Disease*,

https://doi.org/10.1007/978-3-319-92411-3_19

2012, hundreds of animals were already infected when the LSD epidemic was detected in Israel. As a result the veterinary services decided to avoid stamping out. The epidemic was eventually controlled by the use of effective vaccination (Ben-Gera et al. 2015). As opposed to this, in Greece, the LSD epidemics which occurred in 2015 and 2016 continued to spread in areas which were incompletely vaccinated, despite the use of total stamping out of affected herds (Agianniotaki et al. 2017). Similar observations were reported from other countries (AHAW 2017).

A mathematical model, simulating LSD spread in Bulgaria, based on the data of the LSD epidemic, which occurred in Israel during 2012–2013, supports these findings. The model was used to analyse the result of all combinations of three stamping-out possibilities (no stamping out, modified stamping out and total stamping out) and three vaccination statuses (no vaccination, reactive vaccination and pre-emptive vaccination). The results of the model clearly demonstrate that vaccination is the most important factor preventing epidemic spread, while performance of stamping out has only minor influence (AHAW 2016).

Taking the above evidences together, it can be drawn that stamping out can be effective for controlling LSD in secluded areas undergoing frequent and efficient surveillance. Otherwise, effective vaccination is a preferable control method.

References

Agianniotaki EI, Tasioudi KE, Chaintoutis SC, Iliadou P, Mangana-Vougiouka O, Kirtzalidou A, Alexandropoulos T, Sachpatzidis A, Plevraki E, Dovas CI, Chondrokouki E (2017) Lumpy skin disease outbreaks in Greece during 2015–16, implementation of emergency immunization and genetic differentiation between field isolates and vaccine virus strains. Vet Microbiol 201:78–84
AHAW, E.P.o.A.H.a.W. (2015) Scientific opinion on lumpy skin disease. EFSA J 13:3986
AHAW, E.P.o.A.H.a.W. (2016) Urgent advice on lumpy skin disease. EFSA J 14:4573
AHAW, E.P.o.A.H.a.W. (2017) Scientific opinion on lumpy skin disease. EFSA J 15:4773
Babiuk S, Bowden TR, Parkyn G, Dalman B, Manning L, Neufeld J, Embury-Hyatt C, Copps J, Boyle DB (2008) Quantification of lumpy skin disease virus following experimental infection in cattle. Transbound Emerg Dis 55:299–307
Ben-Gera J, Klement E, Khinich E, Stram Y, Shpigel NY (2015) Comparison of the efficacy of Neethling lumpy skin disease virus and x10RM65 sheep-pox live attenuated vaccines for the prevention of lumpy skin disease: the results of a randomized controlled field study. Vaccine 33:4837–4842

Animal Movement Control and Quarantine 20

Eyal Klement

The control of animal movement and animal quarantine may be of an extensive economic burden for affected countries. Therefore, the decision regarding such measures should be considered carefully. As detailed in Epidemiology—chapter in Part I—the average spread velocity of lumpy skin disease in the Balkans was found to be about 7 km/week, and about 99% of the cases were the result of spread of up to 12 km. However, there were occasions of long-distance spread of up to hundreds of kilometres (Mercier et al. 2017). It seems that the virus spreads mainly to adjacent farms by infected vectors but can spread to longer distances via the movement of infected animals. The rationale behind quarantine is to prevent these cases of long-distance virus spread.

In order to control the disease efficiently with a minimum cost, it is necessary to wisely consider the time and distance of the quarantine. Taking into account that the intrinsic incubation period of LSD can take up to 4 weeks in natural conditions (Tuppurainen and Oura 2011), a quarantine period of 4 weeks from the elimination of the last clinical case is reasonable. As stated above, 99% of cases appear within 12 km in the week following detection (i.e. a diameter of 24 km); a safety margin of twice this size (50 km) seems justified for animal movement restriction. There is no official recommendation for quarantine size and period. However (Tuppurainen and Galon 2016) defined a minimum protection and vaccination zone of 50 km in accordance with disease spread evidences.

Nevertheless, in the vast epidemic which occurred in the Balkans during 2015–2016, it was shown that quarantine coupled with stamping out is not sufficient to prevent epidemic spread if not accompanied by effective vaccination (AHAW 2017).

© Springer International Publishing AG, part of Springer Nature 2018   97
E. S. M. Tuppurainen et al., *Lumpy Skin Disease*,
https://doi.org/10.1007/978-3-319-92411-3_20

References

AHAW, E.P.o.A.H.a.W. (2017) Scientific opinion on lumpy skin disease. EFSA J 15:4773

Mercier A, Arsevska E, Bournez L, Bronner A, Calavas D, Cauchard J, Falala S, Caufour P, Tisseuil C, Lefrancois T, Lancelot R (2017) Spread rate of lumpy skin disease in the Balkans, 2015–2016. Transbound Emerg Dis 65(1):240–243

Tuppurainen E, Galon N (2016) Lumpy skin disease: current situation in Europe and neighbouring regions and necessary control measures to halt the spread in south-east Europe. Europe – OIE Regional Commission

Tuppurainen ES, Oura CA (2011) Review: lumpy skin disease: an emerging threat to Europe, the Middle East and Asia. Transbound Emerg Dis 59(1):40–48

Vector Surveillance and Control

21

Yuval Gottlieb

21.1 Vector Surveillance

Field observations and laboratory vector competence tests support the assumption that lumpy skin disease (LSD) virus is mechanically transmitted by arthropods. Various blood-feeding arthropods such as haematophagous flies, mosquitoes and ticks are suspected as potential vectors, but their vectorial capacity has never been tested (see detailed descriptions in the Epidemiology chapter). Given the mechanical transmission mode and the large distribution of LSD from Africa to Europe over diverse climate regions, it is possible that different vectors are found in different areas and also in different farm settings and locations within a small area. Thus, for assessing vectorial capacity and prompting vector control programmes, the recognition and targeting of the local potential vectors are required. For this, local entomological data should be adequately collected and analysed. Such data can be collected by performing wide-taxa surveys, targeting potential vectors that feed on the affected animals. A wide-taxa survey should be planned appropriately to capture taxa with diverse biological and ecological characters (i.e. activity time and biting rates, dispersal, developmental cycle and life span, resting and breeding sites). Planning should include specific taxon traps or collection methodology, adequate trap locations, time of collection and methodology of identification of the collected arthropods (for general guideline, see http://vectormap.si.edu/Project_ESWG.htm). All steps are crucial for obtaining the relevant information on the vector abundance and seasonal dynamics, hence its relevance for vectorial capacity.

In the case of LSD, field observations and experimental results identify ticks (Acari: Ixodida), mosquitoes (Diptera: Culicidae) and stable flies (Diptera: Muscidae) as potential vectors (Tuppurainen et al. 2013; Chihota et al. 2001; Kahana-Sutin et al. 2017, respectively). These diverse vectors obviously have a very different biology and ecology and subsequently collection methodologies. While ticks are collected directly from the host, mosquitoes require dedicated traps composed of black light and CO_2 or water baits. For stable flies, sticky traps and direct collection using a

© Springer International Publishing AG, part of Springer Nature 2018
E. S. M. Tuppurainen et al., *Lumpy Skin Disease*,
https://doi.org/10.1007/978-3-319-92411-3_21

sweeping net could be used. Moreover, different species of the above taxa have different seasonality and breeding needs, which requires various sampling locations and time points. Moreover, assuming other suspected taxa such as horn flies (Diptera: Muscidae), *Culicoides* (Diptera: Ceratopogonidae), black flies (Diptera: Simullidae) and horseflies (Diptera: Tabanidae) which are found in close proximity to cattle should also be taken into consideration when planning wide vector survey.

The main purpose of such a survey is to identify the seasonal abundance of potential vectors of LSDV. Concurrent seasonal abundance of particular vectors with the occurrence of outbreaks may provide insights on vector capacity and enable performance of successive control. The first step is to define the survey area, and for this, mapping of accurate locations of cattle holdings and hotspots of morbidity is needed. Environmental parameters around affected areas, such as elevation, climate, water sources and soil type, should be considered when defining the survey area as they may influence vector spread and abundance. Suggested methodologies for conducting wide-taxa vector survey which include information on suspected LSDV vectors biology, dedicated vector traps, preservation of samples and vector identification can be found in European Food Safety Authority (2017).

In order to effectively control LSDV in affected countries, a long-term surveillance of abundant vectors found in the initial survey should follow. This will enable LSDV detection in the collected arthropods during outbreaks for better risk assessment, as well as accurate and efficient vector control programmes in LSD-affected and at-risk countries.

21.2   Vector Control

Up to date, there is no epidemiological evidence for the effectiveness of vector control in the prevention of LSDV, and the current information on LSDV vectors is scarce. Nevertheless, vector control using pesticides has been done in Bulgaria (EFSA 2017), and it is practised in many farm settings regardless of the diseases. Thus, and because there might be a risk in practising vector control measures, general principles in vector control strategies need to be considered. All blood-feeding vectors are different arthropods with common biological features such as high surface to volume ratio, temperature-dependent developmental time, sensitivity to common pesticides and humid environment requirements for their development and sustainability of young stages.

There are various vector control strategies, each has its recognized limitations and advantages, and therefore the accepted and most effective approach is that of integrated vector control (IVM) which combines different control and prevention strategies each may add or compensate for the disadvantage of the other (more information can be found in http://www.who.int/neglected_diseases/vector_ecology/ivm_concept/en/). The most common strategy for vector control is the use of chemical pesticides directed to the young stages (larvicides) or the adults (adulticides). Most of these are commercially available neurotoxins, which are specified for arthropods under the correct dose and accurate application method. List of available pesticides and their mode of action can be found in http://www.alanwood.net/pesticides/ and other websites alike. Since the major and direct limitation of chemical pesticides is the development of resistance, application should be limited and alternate. Information on resistance in arthropods of interest can be found in http://www.irac-online.org/. Another limitation of pesticides is the potential for secondary toxicity due to residuals of the chemical in the environment. The use of each pesticide should take into account the regulations for pesticide usage, which are specific for each country. Due to these limitations, it is important to consider the accurate time for pesticide applications. Derived from the pest control discipline, control actions should be taken when the pest population reaches the economic threshold (ET) level which is set at the pest density in which management action should be taken to prevent an increasing pest population from reaching the economic injury level (EIL). This is when the cost of loss derived from the damage caused by pest population is equal to the cost of control measures (Pedigo et al. 1986). However, since this is usually unknown in most animal vector-borne diseases and certainly in the case of LSD where the vectors are in question, the time of action should be related to the target vector population dynamics (to be determined according to surveillance data) and precede the predicted time of outbreaks.

Other control strategies are mostly aimed for preventing increase in vector population size (which is directly associated with outbreak risk). These include mechanical and management control. The first is aimed to protect the animal from contact with vectors, mostly using confined housing and nets as well as traps. The second mostly refers to the elimination of vector breeding sites, as in most of the vectors, the majority of the population is found in the juvenile stages. Breeding site varies among taxa, but many potential dipteran vectors can breed in manure, which is the mixture of dung, soil and vegetation. These, however, are less relevant for herds in pasture. Other prevention methods include repellents such as DEET and various oil-based products, as well as pesticides such as permethrin.

Reduction of vector population can also be performed with natural enemies such as microorganisms, predators and parasitoids (DeBach and Rosen 1991). Vector eradication is also a specific taxa methodology and includes the insect sterile technique successfully used for various flies (Dyck et al. 2006) and genetic manipulation of the vectors (Phuc et al. 2007). Constant efforts are also given for reduction of the vector competence (see eliminatedengue.com as an operating methodology example); however, most of these strategies are under development and ahead of the current information on LSDV.

Specific options for control of various taxa of vectors can be found in Mullen and Durden (2009). When planning control programme, risk assessment for any of the control methodologies should also be considered, taking into account the effect on the animals, the farmers and the immediate environment and its inhabitants. Guidelines for pesticide risks can be found in http://www.who.int/whopes/guidelines/en/.

Assuming the vectors of LSD are commonly found in cattle-rearing settings, there are several practices that can be done regardless of the knowledge on the specific vector, especially for confined cattle. In the aim of breaking the vector life cycle, interrupting breeding site methods may be easy to apply by dehydrating manure pits, frequent cultivation of cattle yard and frequent removal or cover of out-of-farm breeding habitats (water source, manure and silage). These practices, coupled with augmentation of natural enemies (e.g. commercially produced parasitoids), will reduce the size of young stage populations. The use of several types of traps which need to be constantly monitored and replaced could be useful to reduce adult numbers and can also help in identifying the potential vectors. In addition, farm practice such quarantine during outbreaks and summer ventilation could also help in expelling adult vectors from the animals. In addition, as mentioned above, the use of pesticides before an expected vector population rises can be used locally for both young and adult stages according to legal permits in the farm environment and on the animals.

References

Chihota CM, Rennie LF, Kitching RP, Mellor PS (2001) Mechanical transmission of lumpy skin disease virus by Aedes aegypti (Diptera: Culicidae). Epidemiol Infect 126:317–321
DeBach P, Rosen D (1991) Biological control by natural enemies. CUP, Cambridge, UK
Dyck VA, Hendrichs J, Robinson AS (2006) Sterile insect technique: principles and practice in area-wide integrated pest management. Springer, Dordrecht
European Food Safety Authority (2017) Lumpy skin disease: I. Data collection and analysis. EFSA J 15:e04773
Kahana-Sutin E, Klement E, Lensky I, Gottlieb Y (2017) High relative abundance of the stable fly Stomoxys calcitrans is associated with lumpy skin disease outbreaks in Israeli dairy farms. Med Vet Entomol 31(2):150–160
Mullen G, Durden L (2009) Medical and veterinary entomology, 2nd edn. Elsevier, Amsterdam
Pedigo LP, Hutchins SH, Higley LG (1986) Economic injury levels in theory and practice. Annu Rev Entomol 31:341–368
Phuc HK, Andreasen MH, Burton RS, Vass C, Epton MJ, Pape G, Fu G, Condon KC, Scaife S, Donnelly CA (2007) Late-acting dominant lethal genetic systems and mosquito control. BMC Biol 5:11
Tuppurainen ES, Lubinga JC, Stoltsz WH, Troskie M, Carpenter ST, Coetzer JA, Venter EH, Oura CA (2013) Mechanical transmission of lumpy skin disease virus by Rhipicephalus appendiculatus male ticks. Epidemiol Infect 141:425–430

Decontamination and Disinfection

22

Eeva S. M. Tuppurainen

Although lumpy skin disease (LSD) is mainly a vector-borne disease, transmission of lumpy skin disease virus (LSDV) is likely to occur also via indirect contact with contaminated environment. Therefore, disinfection of cattle holdings, including both the premises and the environment, during and after an outbreak should be one of the priorities in controlling LSD. Decontamination is particularly important after a stamping-out measure, prior to introduction of the new cattle to the affected holding, regardless if the new animals are vaccinated or not.

LSDV is a large enveloped DNA virus which is remarkably stable between pH 6.6 and 8.6 (Weiss 1968). Laboratory experiments have demonstrated that the virus survives well freezing and thawing with only slightly reduced infectivity (Haig 1957). Thus, LSDV is likely to survive and remain infective in the environment after exposed to below 0 °C temperatures of the winter time. In unclean shaded pens, the virus is surrounded and protected by organic material, and thus, it may persist viable and infectious at least for 6 months, whereas in higher temperatures, the virus is inactivated at 56 °C for 2 h and at 65 °C for 30 min (OIE 2016).

In general, purified LSDV is sensitive for the most of the disinfectants when used in appropriate concentrations. When carrying out disinfection of the laboratory facilities and equipment, it should be taken into consideration that corrosion of metal surfaces may be caused by some disinfectants.

The Food and Agriculture Organization (FAO) of the United Nations provides detailed practical recommendations for proper decontamination of premises, equipment and environment in the Animal Health Manual (FAO 2001). In addition, the Organization for Animal Health (OIE) provides advice on disinfection in the Terrestrial Animal Health Code and other publications (Fotheringham 1995; OIE 2017).

Affected cattle with multiple skin lesions shed scabs and crusts, containing high titres of infectious virus, to the environment (Davies and Otema 1981). The virus remains well protected inside of the scabs, and many disinfectants may not penetrate the organic material without losing their effectiveness. Therefore prior to disinfection, a thorough cleaning of the premises is required in order to achieve effective

© Springer International Publishing AG, part of Springer Nature 2018   103
E. S. M. Tuppurainen et al., *Lumpy Skin Disease*,
https://doi.org/10.1007/978-3-319-92411-3_22

chemical decontamination. Dirt, fat, grease, dung and other surface material must be removed and either burned or buried. All surfaces should be washed by brushing with a soap and detergent solution. The contact time for soap solution should be at least 10 min. LSDV is readily inactivated by most detergents such as sodium dodecyl sulphate or equivalent (FAO 2001).

When carrying out decontamination, buildings with wooden or metallic structures, machinery of mostly metallic components, pipework of various types, water tanks, animal food storage areas and sewage waste should also be included. Hot water with high pressure and steam is effective in cleaning cracks and crevices.

Only after thorough cleaning, the actual disinfection can take place. Many effective disinfectants are toxic both for animal and humans, and therefore, appropriate personal protection must be used by the personnel. Potential effect for the environment should be considered prior to use.

In regard to chlorine compounds, presence of organic matter (such as bovine faeces) has been shown to decrease the levels of active chlorine, common in many disinfectant products. Also incorrect storage and exposure of the disinfectant compound to light may decrease in chlorine concentration in the product (de Oliveira et al. 2011).

For other poxviruses such as the smallpox and vaccinia, several disinfectants have been demonstrated effective, including ethanol (70%), isopropyl alcohol (50%) and active chloride compounds (2.0%) such as sodium (NaClO) (bleach) or calcium hypochlorite ($CaCl_2O_2$). Also formaldehyde (1%), phenol (2%) and iodine compounds are effective (Tanabe and Hotta 1976; de Oliveira et al. 2011).

Aldehydes such as glutaraldehyde are commonly used disinfectants and available by many producers. A 2% dilution (w/v) is effective for 10–30 minutes. In case 40% formaldehyde is used in 1:12 dilution (8% w/v for 10–30 min), it should be taken into consideration that it releases irritating toxic gas.

Compounds with iodine as the active ingredient, a concentration of 0.045% iodine, have been showed to reduce the virus titres of a closely related vaccinia virus by 100% after 1, 5 or 30 min. For example, for a product with active ingredient of 11.25 w/v a required dilution was 1:250 (de Oliveira et al. 2011).

For quaternary ammonium compound products with an active ingredient of 30 w/v, the recommended dilution is 1:2000.

Sodium hydroxide pellets (2%, 20 g/L, caustic soda) with a 10-minute contact time are very effective against viruses but corrosive. Washing soda ($Na_2CO_3 \cdot 10H_2O$) (10%, 100 g/L) should be allowed 30 min contact time and anhydrous sodium carbonate, (Na_2Co_3) (4% solution, 40 g/L for 10 min) 10 min contact time, in the presence of high concentrations of organic material (FAO 2001).

For safe decontamination of clothes and personnel, 2% Virkon® or citric acid powder in 2% solution (2 g diluted in a litre of water) can be used (FAO 2001).

In practise, infected carcasses are often disposed by burying or burning. If required contaminated milk originating from severely infected cows can be treated with disinfectant or buried with the culled animals. Lime (calcium carbonate, dolomite) can be used for disinfection of contaminated soil, litter and manure.

Automatic feeding and drinking machines may be cleaned and disinfected using the same disinfectants, but care should be taken with those machines containing aluminium parts which can be liable to corrosion. In these cases, rinsing with portable water after disinfection will be necessary (Fotheringham 1995).

References

Davies FG, Otema C (1981) Relationships of capripoxviruses found in Kenya with two Middle Eastern strains and some orthopoxviruses. Res Vet Sci 31:253–255

de Oliveira TML, Rehfeld IS, Guedes MIMC et al (2011) Susceptibility of vaccinia virus to chemical disinfectants. Am J Trop Med Hyg 85:152–157. https://doi.org/10.4269/ajtmh.2011.11-0144

FAO (2001) Manual on procedures for disease eradication by stamping out. In: FAO Animal Health Manual http://www.fao.org/docrep/004/Y0660E/Y0660E04.htm. Accessed 4 Jan 2017

Fotheringham VJC (1995) Disinfection of livestock production premises. Rev Sci Tech 14:191–205

Haig DA (1957) Lumpy skin disease. Bull Epizoot Dis Afr 5:421–430

OIE (2016) Lumpy skin disease. In: OIE Terrestrial Animal Health Code. http://web.oie.int/eng/normes/mcode/en_chapitre_1.11.12.htm#rubrique_dermatose_nodulaire_contagieuse. Accessed 27 Aug 2016

OIE (2017) General recommendations on disinfection and disinsection. In: OIE Terrestrial Animal Health Code (Chapter 4.13). http://www.oie.int/index.php?id=169&L=0&htmfile=chapitre_disinfect_disinsect.htm

Tanabe I, Hotta S (1976) Effect of disinfectants on variola virus in cell culture. Appl Environ Microbiol 32:209–212

Weiss KE (1968) Lumpy skin disease virus. Virol Monogr 3:111–131

Active and Passive Surveillance

Eeva S. M. Tuppurainen

Due to its substantial economic impact, lumpy skin disease (LSD) is categorized as a notifiable disease by the World Organization for Animal Health (OIE). An outbreak or use of vaccines against LSD inflicts immediate restrictions to the export of live cattle and some of their products. After an outbreak or after vaccinations have been ceased, affected countries need to establish disease surveillance to restore the freedom-of-disease status. In recently (2014–2017) affected countries in Southeast Europe, the spread of the disease was swiftly halted using harmonized regional vaccination campaigns against LSD. Ideally, the collaboration between affected countries should be continued by conducting the surveillance programmes aiming to regain the disease-free status, also in a regional harmonized manner.

The restrictions and conditions for trade of live cattle and their products are described in the OIE Terrestrial Code, Infection with Lumpy Skin Disease Virus chapter. The OIE gives guidance for the surveillance programmes in the Terrestrial Animal Health Code (OIE 2017), providing detailed advice on how to carry out clinical and laboratory surveillance in the field, such as how often farms should be visited, what samples should be collected and how many animals should be sampled in order to obtain statistically meaningful results.

The European Commission (EC) also requires that affected countries carry out surveillance programmes in order to fulfil the conditions for trade of live cattle and their products. Within the European Union, a LSD-free country under imminent threat of LSD can carry out pre-emptive vaccinations in the whole country or at high-risk zones provided that the country has submitted the Commission a vaccination plan in advance for approval, data for each vaccinated animal is recorded in an electronic cattle ID/movement/vaccination database and an intensified LSD surveillance programme around vaccinated zones is in place (EU 2016/2009 of 15 November 2016). In 2017, only Croatia and Bulgaria have a free-with-vaccination status in regard to LSD.

© Springer International Publishing AG, part of Springer Nature 2018   107
E. S. M. Tuppurainen et al., *Lumpy Skin Disease*,
https://doi.org/10.1007/978-3-319-92411-3_23

The general aims of the surveillance programmes are to demonstrate the absence of the disease, to evaluate the presence or distribution of disease or to enable early detection of infected cases in order to be able to eradicate the disease without delay.

Due to the highly characteristic clinical signs of LSD, active and/or passive clinical surveillance are the most effective tools but should be supported by a laboratory surveillance.

The capability to recognize the clinical signs of LSD and basic understanding on the modes of transmission of the virus, by different stakeholders of the cattle farming industry, are a prerequisite for a successful active and particularly passive clinical surveillance. Awareness campaigns should be targeted to high-level policymakers, veterinary services, official and private field veterinarians, veterinary students, farmers, farm workers, animal traders and their staff as well as slaughterhouse personnel. Enhanced understanding on the characteristics of LSD assists decision-makers to set up a feasible and cost-effective control and eradication plan. Farmers, farm workers or anyone who comes into contact with cattle is high on the priority list to be trained to recognize animals infected with LSD virus. The sooner private or official veterinary officers are alerted, the faster the official disease investigation can be initiated, allowing early implementation of the disease control and surveillance measures. In addition, better understanding on the different modes of transmission of the virus helps farmers to improve the biosecurity in their holdings. Importantly, drivers of the cattle transport vehicles should be included in training in order to prevent shipment of infected cattle showing skin lesions which is a major risk factor for the spread of LSD.

Means to build awareness include training courses, train-a-trainer courses, press releases, information brochures, leaflets and posters. Advertisements through radio and television broadcasts are effective ways to disseminate news throughout the farming community as well as presentations in farmers or cattle association meetings.

Laboratory surveillance is more challenging. Serological surveillance needs to be carefully planned because the timeframe to detect raised antibody levels in affected or vaccinated cattle population is likely to be less than a year after an outbreak or vaccination. Later the immunity will shift to the cell-mediated side (Kitching et al. 1987). Previous studies have shown that the immune status of cattle infected several years ago or that of vaccinated animals cannot be directly related to serum levels of neutralizing antibodies (Weiss 1968; Kitching 1986). In addition, some serological tests, such as serum neutralization assay, may not be sufficiently sensitive to detect mild and old LSDV infections or antibodies in vaccinated animals (Kitching et al. 1987).

As there are not yet any differentiating infected from vaccinated animals (DIVA) vaccines available for LSD, soon after the vaccination campaign, serological surveillance is not useful unless the purpose of the study is to measure the antibody response in vaccinated animals. However, it should be noted that all vaccinated animals may not seroconvert although they would be fully protected. After vaccination, antibodies usually appear within 15 days and reach the highest level 30 days postinoculation (Weiss 1968), but then slowly the antibody levels decrease below detectable levels. In naturally infected animals, antibodies against LSDV can usually be detected from 3 to 6 months postinfection but this time may be longer, and more studies are required to investigate the actual duration of the humoral response in both naturally infected and vaccinated cattle.

Because all members of the *Capripoxvirus* genus cross-react serologically, by using serological methods, it is not possible to differentiate infected from vaccinated animals, even though sheepox- or goatpox-containing vaccines had been used for cattle against LSDV.

Serological surveillance is a highly valuable tool in at-risk countries neighbouring infected regions when there is a need to monitor if unnoticed or unreported outbreaks have occurred at high-risk zones.

In 2017 an antibody detecting ELISA became commercially available for LSD, sheepox and goatpox. ELISA is superior to neutralization assays as it does not require availability of high-containment laboratory facilities or working with live virus and cell cultures. The new test is sufficiently sensitive and can be used in countries affected with LSD and where, instead of vaccination, the main disease eradication measure is a total stamping out. ELISA is also useful in "free-with-vaccination" countries where intensified surveillance zones need to be created around the vaccinated zones. ELISA can be used to test cattle for export or import, although it should ideally be combined with molecular assays to exclude viraemia in tested animals.

Molecular assays are very sensitive and useful for surveillance if slaughter of all infected and in-contact animals is not feasible or affordable in the country, but viraemic asymptomatic animals need to be identified and removed from affected herds, as these animals can serve as a source of infection via blood-feeding vectors. Often in a field situation, vaccination campaigns have been started late, when the disease is already circulating in the region and clinical signs of LSD are detected in vaccinated cattle. If a homologue LSDV-containing vaccine is used, gel-based and real-time PCR methods for differentiation of the vaccine virus from the virulent field strain have been described. If clinical signs are detected in herds vaccinated with a sheepox or goatpox virus-containing vaccine or clinical signs are detected in wild ruminants, species-specific genotyping PCR methods are available. Alternatively sequencing of the appropriate gene sequences such as GPCR or RPO30 can be used. All these methods are described in Diagnostic Tools chapter (Part II, Chap. 16). Due to the cell-mediated immunity against LSDV and low antibody response in some infected or vaccinated animals, development of a cell-based assay would be ideal and is urgently required.

References

Kitching RP (1986) The control of sheep and goat pox. Rev Sci Tech Off Int Epizoot 5:503–511

Kitching RP, Hammond JM, Taylor WP (1987) A single vaccine for the control of capripox infection in sheep and goats. Res Vet Sci 42:53–60

OIE-World Organization for Animal Health (2017). Animal Health Surveillance, Chapter 1.4. In: OIE Terrestrial Animal Health Code. http://www.oie.int/index.php?id=169&L=0& htmfile=chapitre_surveillance_general.htm

Weiss KE (1968) Lumpy skin disease virus. Virol Monogr 3:111–131